93 Springer Series in Solid-State Sciences

Edited by Peter Fulde

Springer Series in Solid-State Sciences

Editors: M. Cardona P. Fulde K. von Klitzing H.-J. Queisser

Managing Editor: H. K. V. Lotsch   Volumes 1–89 are listed at the end of the book

90 **Earlier and Recent Aspects of Superconductivity**
Editors: J. G. Bednorz and K. A. Müller

91 **Electronic Properties of Conjugated Polymers III**
Basic Models and Applications
Editors: H. Kuzmany, M. Mehring, and S. Roth

92 **Physics and Engineering Applications of Magnetism**
Editors: Y. Ishikawa and N. Miura

93 **Quasicrystals**
Editors: T. Fujiwara and T. Ogawa

94 **Electronic Conduction in Oxides**
By N. Tsuda, K. Nasu, A. Yanase, and K. Siratori

95 **Solid State Chemistry**
By L. C. Kimerling

96 **Electron Liquids**
By A. Isihara

97 **Localization and Confinement of Electrons in Semiconductors**
Editors: F. Kuchar, H. Heinrich, and G. Bauer

T. Fujiwara T. Ogawa (Eds.)

Quasicrystals

Proceedings of the 12th Taniguchi Symposium, Shima, Mie Prefecture, Japan, 14–19 November, 1989

With 95 Figures

Springer-Verlag
Berlin Heidelberg New York London
Paris Tokyo Hong Kong Barcelona

Professor Takeo Fujiwara
Department of Applied Physics, The University of Tokyo,
7-3-1 Hongo, Bunkyo-ku, Tokyo 113, Japan

Professor Tohru Ogawa
Institute of Applied Physics, University of Tsukuba, Tsukuba 305, Japan

ISBN-13: 978-3-642-84255-9 e-ISBN-13: 978-3-642-84253-5
DOI: 10.1007/978-3-642-84253-5

Library of Congress Cataloging-in-Publication Data. Taniguchi International Symposium on the Theory of Condensed Matter (12th : 1989 : Shima-chō, Mie-ken, Japan) Quasicrystals : proceedings of the 12th Taniguchi Symposium, Shima, Mie Prefecture, Japan, 14–19 November 1989 / T. Fujiwara, T. Ogawa, eds. p. cm.–(Springer series in solid-state sciences ; 93) Includes bibliographical references and index.
ISBN-13: 978-3-642-84255-9
1. Quasicrystals–Congresses. 2. Condensed matter–Congresses. 3. Crystallography–Congresses. 4. Atomic structure–Congresses. 5. Electronic structure–Congresses. I. Fujiwara, T. (Takeo), 1944– . II. Ogawa, Tōru, 1936– . III. Title. IV. Series. QC173.4.Q36T36 1989 530.4'1–dc20 90-10253

Softcover reprint of the hardcover 1st edition 1990

2154/3140-543210 – Printed on acid-free paper

Preface

This volume contains papers presented at the Twelfth Taniguchi Symposium on the Theory of Condensed Matter, which was held at Kashikojima (in Ise-Shima National Park), Japan, November 14–19, 1989. The general purpose of the Taniguchi Symposia is to encourage important developing, rather than established, fields in condensed matter theory. The topic of the present symposium, *Quasicrystals*, is quite typical.

In 1984, Shechtman, Blech, Gratias and Cahn discovered the icosahedral symmetry of a diffraction pattern and Levine and Steinhardt independently presented the notion of quasicrystals. Before these discoveries, Roger Penrose of Oxford University had invented a space-filling non-periodic tiling, now called *Penrose tiling*. These factors form a new field that had become mathematically viable by the end of 1984, and many important new ideas are still being created. In standard textbooks of solid-state science, the first chapter used to be devoted to symmetry and periodicity in crystals. Now, the textbooks should be revised; quasi-periodicity and its physical properties should be added in several chapters and almost all standard conceptions should be reconsidered. However, the facts that are known about quasiperiodicity are not enough to complete even an introductory chapter of a textbook. Revision should be extended to generalized crystallography, defects, crystal growth, electronic structure, spectral theory and localization, electron transport, spin statistics, etc. These are all topics treated in this volume.

Part I introduces several ideas about quasicrystals and roughly outlines the contents of this volume. Part II is devoted to generalized crystallography and symmetry. The contributions in Part III are about atomic structures and experimental observation. Part IV treats several aspects of random tiling models and perfect quasicrystal models. Parts V and VI discuss physical properties of quasicrystals other than structure, such as electronic structure and spin statistics.

The present symposium is fully supported by the Taniguchi Foundation. The basic idea of the Taniguchi Symposia is, as stated in the speech by Mr. Toyosaburo Taniguchi at the symposium reception, not only to encourage special topics in condensed matter theory, but also to promote international friendship and cooperation. We, all the participants, will bear in mind the memory of the pleasant time we enjoyed in Kashikojima and will continue to strive for international cooperation and mutual understanding.

On behalf of all the participants, we would like to express our sincere gratitude to Mr. Toyosaburo Taniguchi, Prof. Ryogo Kubo, Prof. Junjiro Kanamori and the Taniguchi Foundation for organizing this fruitful and pleasant symposium for a beautiful autumnal week in Japan.

Tokyo, Tsukuba
January 1990

Takeo Fujiwara
Tohru Ogawa

Foreword

Speech by Toyosaburo Taniguchi

Dr. Kubo, Distinguished Guests, and Friends,

I am very happy, pleased and honored to be here this evening with so many distinguished guests, friends, and scholars from within this country and from different parts of the world. As a typical Japanese industrialist, I deem it a privilege to be given this opportunity of meeting world-famous scholars in this field. The Taniguchi Foundation wishes to extend a warm and sincere welcome to the many participants in the Twelfth International Symposium on the Theory of Condensed Matter. This symposium series was inaugurated 11 years ago through the great efforts of Dr. Ryogo Kubo, who is gracing us today with his presence.

The conception of the symposium is unique, in that Japanese and foreign participants, being limited to a number suitable for a discussion group, live together in a close-knit community, albeit for a limited period of time. The type of setup and background to the symposium will help to strengthen ties of friendship and understanding with their colleagues on a most personal basis. It is easier said than done to realize the ideal of international friendship and understanding in a world growing steadily smaller. We are deeply indebted to Dr. Kubo and his associates for their efforts, which were crucial in making this particular symposium possible. In this sense, any words of praise should be directed to all those who have lent their support to this program.

So far, I have met in person a total of 87 participants in this Division from 18 foreign countries and 176 participants from Japan. Whenever I meet the participants in our symposia, both young and old, I am deeply impressed by the unselfish and sincere dedication they display in pursuing their mission. I hope and trust that even after ascending to academic fame within, say, the next 10, 20, or 30 years, they will join forces and help to forge closer bonds of friendship and cooperation in a manner that will make a major contribution not only to academia but also to permanent world peace and to the welfare of mankind.

In conclusion, we sincerely trust that all the participants, both Japanese and foreign, will return home bearing heartwarming memories of the symposium and of the pleasant times we have enjoyed together.

Thank you.

Participants in the Taniguchi International Symposium

November 14–19, 1989 (at Kashikojima, Ise-Shima National Park, Japan)

Seated (left to right)
K. Hiraga, C.L. Henley, T. Ninomiya, M.V. Jarić, N.D. Mermin, R. Kubo, P.J. Steinhardt, T. Fujiwara, T. Ogawa, Y. Ishii, H. Hiramoto

Standing (left to right)
A. Inoue, D. Gratias, T. Janssen, T. Tokihiro, A. Yamamoto, K. Sekimoto, M. Kohmoto, M. Widom, H. Tsunetsugu, Y. Okabe, J.E.S. Socolar

Contents

Part IV Quasiperiodic Lattices and Random Tiling

Part V Electronic Structure

Part I

Introduction

Introduction to Quasicrystals

T. Fujiwara

Department of Applied Physics, University of Tokyo,
Bunkyo-ku, Tokyo 113, Japan

Abstract: The discovery by Shechtman, Blech, Gratias and Cahn of sharp diffraction patterns of icosahedral symmetry in Al-Mn alloy rapidly quenched from the melt broke ground in a new field of physics, chemistry, materials science, *etc*. A short introductory review on quasicrystals and an outline of this volume are given.

1. Discovery of non-crystalline rotational symmetry

For many decades, it had been believed beyond any shadow of doubt that the five-fold rotational symmetry cannot exist in equilibrate condensed phases. Mineralogists have discovered many alloy materials containing icosahedral units either isolated or linked, but these materials are globally of crystalline rotational symmetry. Several metallic glasses contain specific local icosahedral symmetry (*i.e.* the icosahedral bond orientational order).[1] In the meantime, our understanding has been widening in related fields.

An icosahedron is locally the most densely packed arrangement but an icosahedral rotational symmetry is contradictory to the translational periodicity. Therefore, it was thought that a long range order of icosahedral symmetry was unlikely to be found in realistic condensed matter, though a two-dimensional pentagonal tiling, called *Penrose tiling*, was found by R.Penrose in 1974.[2] Since the invention of Penrose tiling, de Bruijn [3], Mackay [4], Kramer [5] and Kramer and Neri [6] have studied two- and three-dimensional tilings and their mathematical properties. J.F.Sadoc gave an introductory lecture on the Penrose tilings on the occasion of the fifth Taniguchi International Symposium on Topological Disorder in Condensed Matter in 1982, which was the first time the author saw the Penrose tiling.[7]

The discovery by Shechtman, Blech, Gratias and Cahn [8] of sharp diffraction patterns of icosahedral symmetry (Fig.1) in Al-Mn alloy rapidly quenched from the melt broke ground in a new field of physics, chemistry, materials science, *etc*. The icosahedral phase in rapidly solidified Al-Mn alloy is metastable and remarkably resistant to crystallization up to 350°C. Independently of the discovery by Shechtman *et al.*, Levine and Steinhardt [9] generalized the notion of a crystal in order to include a quasiperiodic translational order and gave it the name *quasicrystal*. Quasicrystals may be specified by two fundamental characteristics, *i.e.* long range quasiperiodic translational order and long range orientational order of crystallographically forbidden symmetry. The former aspect distinguishes quasicrystals from glasses and the latter from incommensurate crystals. However, readers should notice that the ***icosahedral glass*** model [10] and the ***random tiling*** model [11] explain some observed properties of the icosahedral phase.

There are two classes of quasicrystals: metastable quasicrystals (AlMn [8], AlMnSi [12], Mg(Al,Zn) [13], *etc.*) and stable quasicrystals (T2 phase of AlCuLi (Al_6CuLi_3) [14] and recently discovered AlCuFe [15] and AlCuRu [16]). The alloy series AlCuTM (TM=transition metal element) shows quite various phases such as glasses, metastable

Springer Series in Solid-State Sciences, Vol. 93 Quasicrystals
Editors: T. Fujiwara · T. Ogawa

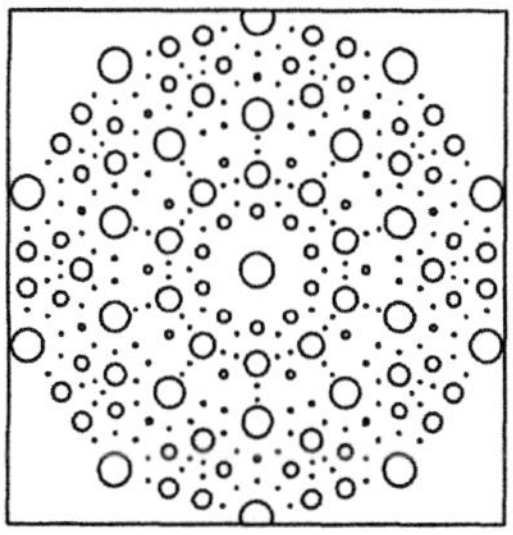

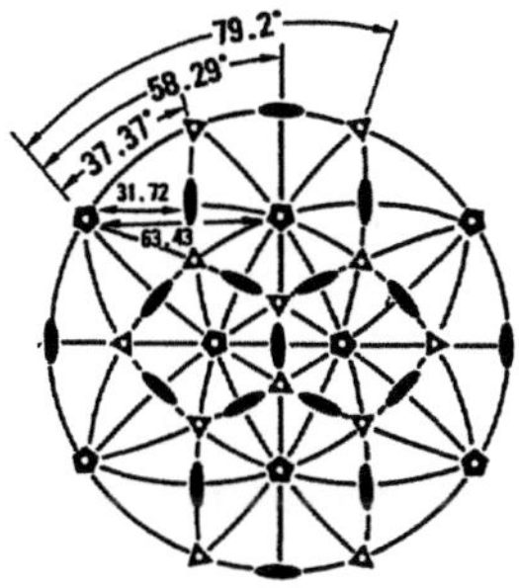

Fig.1 Diffraction pattern of five-fold symmetry (area proportional to the intensity) and stereogram of the icosahedral group.

and stable quasicrystals and crystals.[17] In AlMn [18], AlFe [19] and AlCuCo [17], a decagonal phase has been found whose diffraction pattern is of ten-fold rotational symmetry and which has a one-dimensional translational periodicity along the ten-fold rotational axis. Two-dimensional octagonal quasicrystals of eight-fold rotational symmetry have been obtained in CrNiSi and VNiSi.[20]

Two different types of large *facets* have been observed; one is a rhombic triacontahedron in Al_6CuLi_3 [21] and the other is a pentagonal dodecahedron in AlCuFe [15] and GaMgZn [22]. The equilibrium shape is determined by minimizing the surface free energy and the faceting shapes were considered to be a natural outcome of a long range translational and bond orientational order of a periodic lattice. However, it has been shown that the long range positional order is not essential for facet formation [23] and also the observed faceting shape is a result of an incorporation of attractive and repulsive interactions [24].

2. Quasiperiodic lattice, random tiling and icosahedral glass

Structures of quasicrystals can be generated by *inflation-deflation* operation [3], utilization of *matching rules* [3], the *grid method* [3], *(strip) projection method* [5, 25], *cut projection method* [26] or *generalized dual method* [27]. These methods are closely related to each other.

In the strip projection method, the vertices of the d-dimensional quasiperiodic lattice are obtained by projection of lattice points in a higher-dimensional hypercubic lattice in the n-dimensional space $\mathbf{E}^n$. Consider an orthogonal decomposition of $\mathbf{E}^n$,

$$\mathbf{E}^n = \mathbf{E}^{\perp} + \mathbf{E}^{\parallel}, \tag{1}$$

where the subspace $\mathbf{E}^{\parallel}$ is the d-dimensional physical space. Let $\mathbf{W}$ be the projection of a unit 'hypercube' onto $\mathbf{E}^{\perp}$ (perpendicular-space), which we call the *window*. For any $\mathbf{x}_0$ in $\mathbf{E}^{\parallel}$, let $\mathbf{W}_0$ denote $\mathbf{W} + \mathbf{x}_0$. Once a lattice point in $\mathbf{E}^n$ is located inside $\mathbf{W}_0$, the projection of this lattice point onto $\mathbf{E}^{\parallel}$ should be selected as a vertex in the subspace $\mathbf{E}^{\parallel}$. When the 'slope' of the hyperplane $\mathbf{E}^{\parallel}$ is rational to the hypercubic lattice in $\mathbf{E}^n$, the projected lattice in $\mathbf{E}^{\parallel}$ is periodic. On the other hand, if the 'slope' of the hyperplane $\mathbf{E}^{\parallel}$ is irrational, the projected lattice is quasiperiodic. The n-dimensional hypercubic lattice, the directions of the two orthogonal subspaces $\mathbf{E}^{\perp}$ and $\mathbf{E}^{\parallel}$, and the window $\mathbf{W}$ must be invariant under operations of the non-crystalline group of the quasicrystal.

A one-dimensional quasiperiodic lattice model by the projection from the two-dimensional square lattice is shown in Fig.2. The resulting one-dimensional quasiperi-

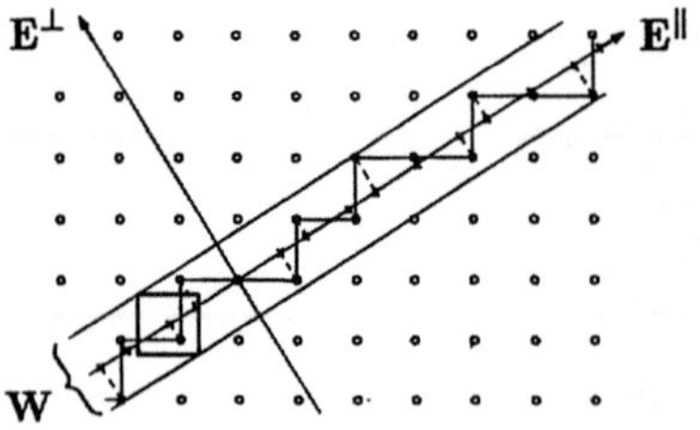

Fig.2 One-dimensional quasiperiodic lattice (cross marks on the $\mathbf{E}^\parallel$-axis) by projection.

odic lattice is called the *Fibonacci lattice*, when the $\mathbf{E}^\parallel$ axis is chosen to be oriented at an angle θ to the x-axis, where $\tan\theta = 1/\tau$ and τ is the golden mean $(\sqrt{5}+1)/2$. A lattice point distribution function $\rho(\mathbf{x})$ is given by

$$\rho(\mathbf{x}) = \rho(x^\parallel, x^\perp) = \sum_{j,\ell} \delta(x^\parallel - j\cos\theta - \ell\sin\theta) \times \delta(x^\perp + j\sin\theta - \ell\cos\theta), \qquad (2)$$

and that of projected lattice on $\mathbf{E}^\parallel$ is

$$\rho_0(x^\parallel) = \int_{-\infty}^{\infty} \mathrm{d}x^\perp \rho(x^\parallel, x^\perp) \times W(x^\perp), \qquad (3)$$

where the window $\mathbf{W}$ is expressed by a function

$$W(x^\perp) = \begin{cases} 1, & \text{if } 2|x^\perp| < \cos\theta + \sin\theta \,; \\ 0, & \text{otherwise.} \end{cases}$$

Therefore, the diffraction pattern of the one-dimensional Fibonacci lattice can be determined by a structure factor

$$f(k) = \int_{-\infty}^{\infty} \mathrm{d}x^\parallel \mathrm{e}^{-ikx^\parallel} \rho_0(x^\parallel) \sim \sum_{m,n=-\infty}^{\infty} \frac{\sin k^\perp w}{k^\perp w} \delta(k - 2\pi(m\sin\theta + n\cos\theta)), \qquad (4)$$

where $w = \frac{1}{2}(\cos\theta + \sin\theta)$ and $k^\perp = 2\pi(m\cos\theta - n\sin\theta)$. This result can be easily generalized to higher dimensional cases and the Bragg spots are generally given as a countable infinite number of densely distributed δ-functions.[9, 28]

The concept of *local isomorphism* is quite important for classifying several quasiperiodic lattices. We say two quasiperiodic lattices are locally isomorphic "if and only if given point P in either tiling and any finite distance d, there exists a pure translation of the other tiling that causes the two to coincide everywhere in a circle of diameter d about P".[29] In particular, if the quasiperiodic lattice is governed by a certain inflation-deflation rule, it is called the *Penrose local isomorphism.*

The two-dimensional pentagonal quasicrystal can be obtained by a projection of the five-dimensional hypercubic lattice and the three-dimensional icosahedral quasicrystal by that of the six-dimensional one. The three-dimensional icosahedral quasicrystals are constructed by two rhombohedral units, prolate and oblate. These are based on a triple of edge vectors of an icosahedron. However, in the icosahedral quasicrystal of the Penrose local isomorphism class, four types of unit cells appear with a uniquely defined matching rule: a triacontahedron, an icosahedron, a dodecahedron and a prolate rhombohedron with volumes in the ratios $10\tau : 5\tau : 2\tau : 1$.[29] The former three types of unit cells can be divided into several prolate and oblate rhombohedra but not uniquely.

Other models of the icosahedral phase competing against the quasiperiodic lattice model are the icosahedral glass model [10] and the random tiling model [11]. In the icosahedral glass model, icosahedral cluster units are randomly aggregated. The rule for constructing the aggregation is to join vertex-to-vertex, edge-to-edge, or face-to-face, with common orientations. The random tiling model is a filling of space by tiles without overlaps and gaps. This model may be constructed by the projection method from a distorted or fluctuating higher dimensional lattice, and gains entropy through this fluctuation. Both of these two different models show, guaranteed by their construction rules, the long range bond orientational order and a diffraction pattern of icosahedral symmetry, though diffraction spots have finite widths in the three-dimensional icosahedral glass model or diffuse background in the three-dimensional random tiling model.

3. Phonon and phason

The mass density $\rho(\mathbf{r})$ can be expanded in a Fourier series,

$$\rho(\mathbf{r}) = \sum_{\mathbf{G}_n} \rho_{\mathbf{G}_n} \exp i\mathbf{G}_n \cdot \mathbf{r} = \sum_{\mathbf{G}_n} |\rho_{\mathbf{G}_n}| \exp i(\mathbf{G}_n \cdot \mathbf{r} + \phi_{\mathbf{G}_n}), \tag{5}$$

where $\mathbf{G}_n$ is the reciprocal vector. Here we are explaining the vertex model of icosahedral quasicrystals. In the vertex model, $\{\rho_{\mathbf{G}_n}\}$ consists of six $\mathbf{G}_n$ corresponding to the position vectors of vertices of an icosahedron shown in Fig.3. The quantities $\theta_n = \mathbf{G}_n \cdot \mathbf{r}$ form a hypercubic lattice in a six-dimensional space, because $\rho(\mathbf{r})$ is invariant under the translation $\theta_i \to \theta_i + 2\pi\ell$.[30] Examples of symmetry operations are

$$\begin{aligned}
C_5 : \mathbf{G}_5 \to \mathbf{G}_5, \quad & \mathbf{G}_0 \to \mathbf{G}_1 \to \mathbf{G}_2 \to \mathbf{G}_3 \to \mathbf{G}_4 \to \mathbf{G}_0, \\
: \theta_5 \to \theta_5, \quad & \theta_0 \to \theta_1 \to \theta_2 \to \theta_3 \to \theta_4 \to \theta_0, \\
C_3 : \theta_1 \to \theta_2 \to \theta_5 \to \theta_1, \quad & \theta_3 \to \theta_0 \to -\theta_4 \to \theta_3, \\
C_2 : \theta_2 \leftrightarrow \theta_3, \quad \theta_0 \leftrightarrow -\theta_5, \quad & \theta_1 \to -\theta_1, \quad \theta_4 \to -\theta_4.
\end{aligned} \tag{6}$$

The character table of the icosahedral group I is given in Table 1 and there are five irreducible representations, $\Gamma_1 \sim \Gamma_5$. The representation Δ of the basis set of the displacements $\{\delta\theta_i\}$ is very important and is reducible into two irreducible representations:

Table 1: Character table of the icosahedral group I.

	E	$12C_5$	$12C_5^2$	$20C_3$	$15C_2$
Γ_1	1	1	1	1	1
Γ_3	3	τ	$-\tau^{-1}$	0	-1
$\Gamma_{3'}$	3	$-\tau^{-1}$	τ	0	-1
Γ_4	4	-1	-1	1	0
Γ_5	5	0	0	-1	1
Δ	6	1	1	0	-2

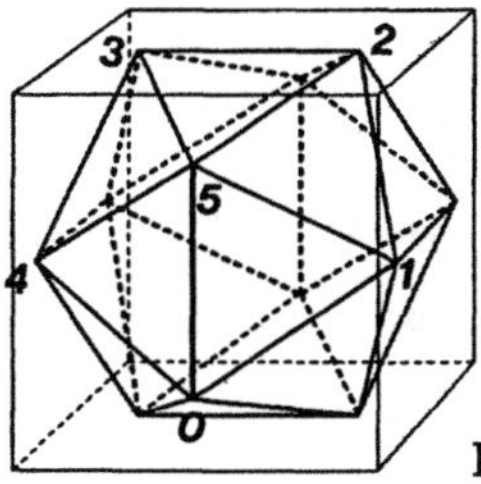

$\mathbf{G}_0 = G(1,0,\tau), \mathbf{G}_1 = G(0,\tau,1), \mathbf{G}_2 = G(-\tau,1,0),$
$\mathbf{G}_3 = G(-\tau,-1,0), \mathbf{G}_4 = G(0,-\tau,1), \mathbf{G}_5 = G(-1,0,\tau)$

Fig.3 Reciprocal vectors $\mathbf{G}_n$ of icosahedral quasicrystals.

$$\Delta = \Gamma_3 \oplus \Gamma_{3'}. \tag{7}$$

The representation Γ_3 is just that of the three translational vectors. A set of reciprocal vectors $\{\mathbf{G}_n\}$ transforms under the representation Γ_3. The set of the complementary vectors

$$\mathbf{G}_5^\perp = -\mathbf{G}_5, \quad \mathbf{G}_n^\perp = \mathbf{G}_{<3n>_5}, n = 0 \sim 4, \tag{8}$$

transforms under the representation $\Gamma_{3'}$, where $< 3n >_5 = 3n$ mod 5. Vectors $\mathbf{G}_n$ and $\mathbf{G}_n^\perp$ satisfy an orthogonality relation. Furthermore, the vectors $(\mathbf{G}_n, \mathbf{G}_n^\perp)$ form a set of six-dimensional orthogonal bases. The scalar variables $\phi_{\mathbf{G}_n}$ are therefore written as

$$\phi_{\mathbf{G}_n} = \mathbf{G}_n \cdot \mathbf{u} + \mathbf{G}_n^\perp \cdot \mathbf{w}, \tag{9}$$

where $\mathbf{u}$ and $\mathbf{w}$ are displacement vectors in the physical $\mathbf{E}^\parallel$ and complementary space $\mathbf{E}^\perp$, respectively. When the displacement vectors $\mathbf{u}$ and $\mathbf{w}$ depend upon the position vector $\mathbf{r}$, $\mathbf{u}(\mathbf{r})$ and $\mathbf{w}(\mathbf{r})$ represent the phonon and phason displacement vectors. Here the name *phason* is borrowed from incommensurate crystals. The phonon displacement $\mathbf{u}(\mathbf{r})$ modulates distances between nearest lattice points, because it causes shifts of the n-dimensional lattice points only within the window. Conversely, by the phason displacement $\mathbf{w}(\mathbf{r})$, some lattice points in the n-dimensional hypercubic lattice move out of the window and some others move into it. This movement rearranges lattice points or the edges of tiles, which can be clearly seen in the grid method.[30, 31] Figure 4 shows the modulation of grids and the corresponding tiling. As can be seen, the phason displacement corresponds to a rearrangement of edges of tiles, or an introduction of the forbidden tiling.

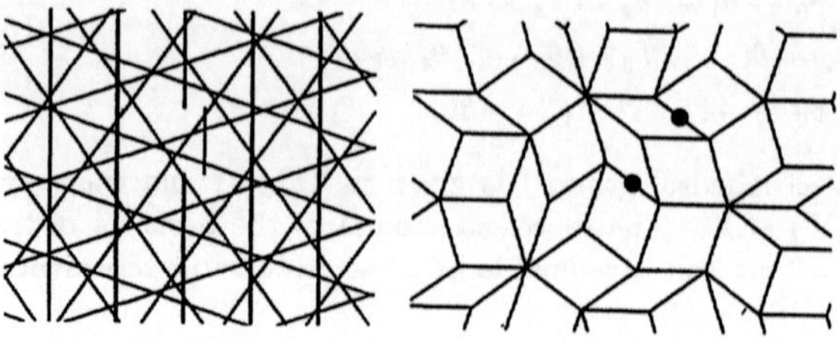

Fig.4 Modulation of grids and corresponding phason. Dots indicate mismatch of tiling.

General Bragg spots may be specified by reciprocal vectors

$$\mathbf{G} = \frac{\pi}{a} \sum_{n=0}^{5} m_n \mathbf{G}_n / (G\sqrt{1+\tau^2}), \tag{10}$$

where m_n are integers and a represents an edge length of the rhombohedral unit cell. The reciprocal lattice point $\mathbf{G}$ is associated with a unique partner in the complementary space

$$\mathbf{G}^\perp = \frac{\pi}{a} \sum_{n=0}^{5} m_n \mathbf{G}_n^\perp / (G\sqrt{1+\tau^2}). \tag{11}$$

These two associated vectors $\mathbf{G}$ and $\mathbf{G}^\perp$ are already seen in Eq.(4), *i.e.*

$$k = 2\pi(m\sin\theta + n\cos\theta) \rightarrow \mathbf{G}, \quad \text{and} \quad k^\perp = 2\pi(m\cos\theta - n\sin\theta) \rightarrow \mathbf{G}^\perp. \tag{12}$$

Therefore, the scattering intensity of the Bragg spots $I(\mathbf{G})$ is proportional to

$$|\int_{\mathbf{W}} d\mathbf{x}^\perp e^{-i\mathbf{G}^\perp \cdot \mathbf{x}^\perp}|^2,$$

where the integral is restricted within the window $\mathbf{W}$ of $\mathbf{E}^{\perp}$-space. The Bragg spots are indexed by a set of integers $(m_5, m_0, m_1, m_2, m_3, m_4)$.[32]

The phonon causes finite widths of diffraction spots through the Debye-Waller factor. As for the phason contributions, if the phason strain is spatially uniform (*linear phason strain* $\mathbf{M}$ where $\mathbf{w}(\mathbf{r}) = \mathbf{M} \cdot \mathbf{r}$), the Bragg spots systematically shift from $\mathbf{G}$ to $\mathbf{G} + \mathbf{G}^{\perp} \cdot \mathbf{M}$, which is observed in several diffraction experiments.[33] Quenched random phasons contribute also to the finite widths of diffraction spots. When the quenched phason is a random Gaussian variable, the widths of the spots are proportional to $|\mathbf{G}^{\perp}|^{2/\alpha}$.[33, 34]

The most serious difficulty of the icosahedral glass model is the systematic shifts and finite widths of the diffraction spots mentioned above. The widths of the diffraction spots in the icosahedral glass model are proportional to $|\mathbf{G}^{\perp}|^2$ ($\alpha = 1$). However, the observed widths of the diffraction spots are proportional to $|\mathbf{G}^{\perp}|$ ($\alpha = 2$) in icosahedral Al_6Li_3Cu.

The idea of the phason is useful even in decagonal (or icosahedral) glass models, which can be considered as the phase modulated by condensed long and short wavelength phasons. In these glass models, assuming that the construction rule is to join edge-to-edge (face-to-face), the coordinate of each aggregated decagonal (icosahedral) unit can be uniquely expressed as $\sum_{i=0}^{n-1} m_i \mathbf{e}_i^{\parallel}$, where $\{\frac{1}{2}\mathbf{e}_i^{\parallel}\}$ is the position vector of a midpoint of an edge (a center of a face) from the center of an aggregated unit. The associated vector field (phason field) defined by $\sum_{i=0}^{n-1} m_i \mathbf{e}_i^{\perp}$ in the complementary space can be a measure of randomness and homogeneity of packing.[35]

The random tiling model can be considered as an icosahedral quasiperiodic lattice model with highly condensed short wavelength phasons and is stabilized by configurational entropy at higher temperatures. [11]

The elastic energy contributed by phonon and phason strains can be easily written down phenomenologically from symmetry considerations.[31] Several complicated crystals with large unit cells can be regarded as lower symmetric modulated phases of icosahedral quasicrystals where the icosahedral symmetry is broken by mode locking of phason strain.[36]

4. Electronic structure

There are three different types of energy spectra in general: *absolutely continuous*, *point* and *singular continuous*.[37] The absolutely continuous spectrum is that with the spectral measure $\mathrm{d}\mu = n(E)\mathrm{d}E$ and has a well-defined smooth density of states $n(E)$ (except a finite number of well-separated singularities). The point spectrum is that whose spectral measure is a set of δ-functions defined on a countable number of points. The resulting integrated density of states is a function with finite steps on a countable number of points. The remaining is the singular continuous spectrum. Its integrated density of states is a continuous function which is monotonically increasing on a set of points with zero Lebesgue measure. An example of the integrated density of states for a singular continuous spectrum is a Devil's staircase, shown in Fig.5. Without any particle-particle interaction, the energy spectrum of the Schrödinger equation is absolutely continuous in a periodic system, corresponding to extended Bloch-type wavefunctions with definite wave vectors. The randomness causes the Anderson localization [38] and the dense point spectrum, whose wavefunctions are exponentially localized.

One important example of one-dimensional quasiperiodic systems is the Harper equation, which is a discrete Schrödinger equation in a two-dimensional square lattice with an applied magnetic field.[39] Choosing the Landau gauge $\mathbf{A} = H(0, x, 0)$, we can obtain a tight-binding Schrödinger equation as

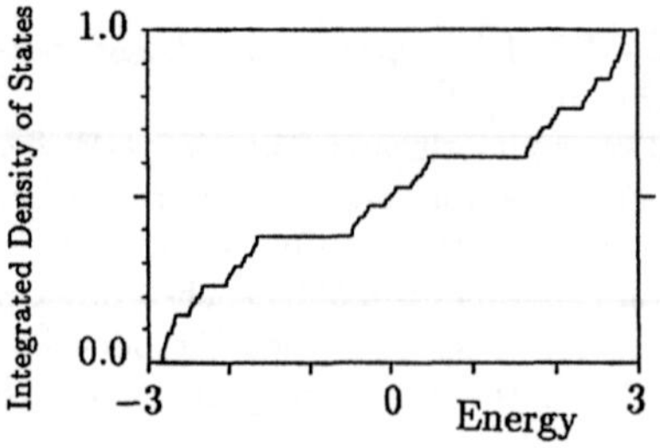

Fig.5 The integrated density of states in a transfer Fibonacci lattice. The transfer parameters are chosen as $t_\tau = 1, t_1 = 2$.

$$t[\psi(x+a,y)+\psi(x-a,y)+\mathrm{e}^{-ieHax/\hbar c}\psi(x,y+a) \\ +\mathrm{e}^{ieHax/\hbar c}\psi(x,y-a)] = E\psi(x,y), \tag{13}$$

where $x = ma$, $y = na$. This equation is periodic in the y-direction. A solution can be found in the form $\psi(ma, na) = \mathrm{e}^{inq}u(m)$ and Eq.(13) is rewritten as

$$u(m+1)+u(m-1)+2\cos(2\pi m\sigma - q)u(m) = \epsilon u(m), \tag{14}$$

where

$$\sigma = a^2H/\frac{ch}{e}, \quad \text{and} \quad \epsilon = \frac{E}{t}.$$

The resulting equation (14) or a more general one (14′) is called the Harper equation:

$$u(m+1)+u(m-1)+2V\cos(2\pi m\sigma - q)u(m) = \epsilon u(m). \tag{14′}$$

If the constant parameter σ is an irrational number, the system is quasiperiodic. We must notice that the one-dimensional system (14′) is not really equivalent to our one-dimensional quasicrystal, because no site has an equivalent site with an identical local environment. The Harper equation has an important self-duality property between itself and the Fourier transformed one. For $V > 1$ and almost every irrational number σ, the energy spectrum cannot be absolutely continuous.[40] (For a rational number σ, the states are of course extended Bloch-type and the spectrum is absolutely continuous.) If $V > 1$ and σ is a Liouville number, the energy spectrum is singular continuous.[40] If V is large enough and σ is a Roth number under some additional condition, the energy spectrum is pure point.[41] A Liouville number is an irrational number which can be approximated by an appropriate rational number.[42] A Roth number is the most "irrational" number and the class of Roth numbers is of full Lebesgue measure.[42] For sufficiently small V, the Harper equation has an absolutely continuous spectrum.[41] Therefore, we can expect the metal-insulator transition to depend upon the potential strength V, irrespective of the irrational number σ (*i.e.* for almost every irrational number σ). It should be mentioned that the golden mean τ is expressed by a continued fraction as

$$\tau = 1 + \cfrac{1}{1+\cfrac{1}{1+\cfrac{1}{1+....}}}$$

and is a Roth number.

Once we replace the continuous potential $V_c(\sigma m) = 2V\cos(2\pi m\sigma - q)$ by a discontinuous one $V_d(\sigma m)$, where

$$V_d(x) = \begin{cases} V, & \text{for } -\sigma \le x \le 0\,; \\ -V, & \text{for } 0 < x < 1-\sigma, \end{cases} \tag{15}$$

and $V_d(x+1) = V_d(x)$, the situation changes quite dramatically. When σ is a rational number, the system is periodic. When σ is irrational, the system is quasiperiodic and it is possible to find, within a finite distance, another region with the same local environment. If σ is equal to $1/\tau = \tau - 1$, the system is just the Fibonacci lattice.[43] Recently, it was rigorously proven that the energy spectrum of the Fibonacci lattice is singular continuous, even if the potential strength is infinitesimal.[44]

There have been several attempts to determine the energy spectrum and the wave-functions in one- and higher-dimensional quasicrystals.[45, 46, 47]

5. Outline of this volume

In the present article, I have tried to give a short introduction to quasicrystals. In this section, I provide an outline of this volume, which is divided into five parts. (Parts II-VI)

In Part II, T.Ogawa discusses in the first two papers a global feature of two-dimensional triangular networks and derives harmonic polynomials of icosahedral symmetry. N.D.Mermin formulates the generalized crystallography in two- and three-dimensions; his trick is to develop the subject entirely in the wavevector space.

Part III is devoted to atomic structures in quasicrystals. C.L.Henley provides a wide and general view of the atomic structures from the theoretical side. M.V.Jarić shows an example of solving the phase problem for determining the atomic mass density function in AlCuLi alloy. A.Yamamoto determines ideal atomic structures of AlMn and AlCuLi quasicrystals. The acceptance domains for all types of atoms are given and the atomic positions of resulting higher crystalline approximants are tabulated. K.Hiraga shows beautiful high-resolution images of local atomic structures of AlCuLi observed with a 400kV electron microscope and explains them.

Part IV contains several papers mainly concerned with the perfect quasiperiodic lattice model and the random tiling model. A.Inoue reviews experimental results of the thermodynamical stability of a series of important alloys AlCuTM and AlNiTM. P.J.Steinhardt discusses a perfect growing quasiperiodic lattice and an important idea, *i.e.* the physical realizability (*restorability* and *growability*). He also discusses a unique role of a defect (decapot) in growing a perfect lattice and mentions local isomorphism classes of the physical realizability. J.E.S.Socolar extends the notion of a matching rule to a weak matching rule which makes a lattice model more realistic, widely applicable and fruitful. M.Widom gives a review of the random tiling model and criticizes the perfect quasiperiodic lattice models on the basis of several experimental results. K.Sekimoto discusses freezing phasons in a macroscopic model from a random tiling picture. Y.Ishii talks about a structural modulation by phasons and a structural stability in some parameter spaces of the elastic energy. It should be mentioned here that scanning tunneling microscopy recently provided a perfect quasiperiodic lattice image of a decagonal quasicrystal AlCuCo on scale as small as 60Å [48], which is a powerful experiment, complementary to the high-resolution electron microscopy shown by K.Hiraga. The coexistence of the perfect icosahedral phase and a large variety of crystalline approximants with large unit cell sizes has been found in Ga-Mg-Zn. [49]

Part V deals with the electronic structures. T.Ninomiya discusses an intuitive picture of the energy spectra and wavefunctions in quasiperiodic lattices. T.Janssen talks about a scale invariance and electron or phonon spectra in well-defined aperiodic lattices. M.Kohmoto gives a review of the general idea of a multifractal method and applies it to exotic energy spectra and wavefunctions of the Fibonacci lattice and the Harper equation. H.Hiramoto shows several examples of pure and mixed energy spectra, Anderson localization and the mobility edges. He also discusses the effect of particle-particle interaction on the energy spectrum. T.Tokihiro presents models with long range hierarchical interactions which can be solved exactly with a self-similar property. H.Tsunetsugu calculates the electric conductance of two-dimensional Penrose lattices and shows, by a finite size scaling technique, that the several wavefunctions decay with a power law. T.Fujiwara discusses the electronic structures in realistic three-dimensional quasicrystals on the basis of those in corresponding crystalline approximants and proposes a mechanism for structural stabilization.

Part VI is a paper by Y.Okabe. He gives a comprehensive Monte Carlo study of ferromagnetic and antiferromagnetic Ising spin systems in two- and three-dimensional Penrose lattices and mentions the universality class of the phase transition.

I hope that readers will obtain from this volume a general idea of the present understanding of quasicrystals. The following may also be a great help for readers wishing to have an overview of quasicrystals.

The Physics of Quasicrystals,
edited by P. J. Steinhardt and S.Ostlund (World Scientific, Singapore 1987).

Introduction to Quasicrystals,
edited by M. V. Jarić (Academic Press, Orlando, FL 1988).

Introduction to the Mathematics of Quasicrystals,
edited by M. V. Jarić (Academic Press, Orlando, FL 1989).

Extended Icosahedral Structures,
edited by M. V. Jarić and D. Gratias (Academic Press, Orlando, FL 1989).

References

1 P.J.Steinhardt, D.R.Nelson and M.Ronchetti, Phys. Rev. B **28**, 784 (1983).
2 R.Penrose, Bull. Inst. Math. Appl. **10**, 266 (1974); M.Gardner, Sci. Am. **236**, 110 (1977).
3 N.G. de Bruijn, Ned. Akad. Weten. Proc. Ser. A **43**, 27 (1981); **43**, 39 (1981); **43**, 53 (1981).
4 A.L.Mackay, Physica **114A**, 609 (1982).
5 P.Kramer, Acta Cryst. A**38**, 257 (1982).
6 P.Kramer and R.Neri, Acta Cryst. A**40**, 580 (1984).
7 R.Mosseri and J.F.Sadoc, *The Structure of Non-crystalline Materials* (Taylor and Francis, London, 1982) p.137.
8 D.Shechtman, I.Blech, D.Gratias and J.W.Cahn, Phys. Rev. Lett. **53**, 1951 (1984).
9 D.Levine and P.J.Steinhardt, Phys. Rev. Lett. **53**, 2477 (1984).
10 D.Shechtman and I.A.Blech, Metall. Trans. **16A**, 1005 (1985); P.W.Stephens and A.I.Goldman, Phys. Rev. Lett. **56**, 1168 (1986).
11 C.L.Henley, J. Phys. A **21** 1649 (1988); M.Widom, D.P.Deng and C.L.Henley, Phys.Rev.Lett. **63**, 310 (1989); K.H.Strandburg, L.H.Tang and M.V.Jarić, Phys.Rev.Lett. **63**, 314 (1989).
12 C.H.Chen and H.S.Chen, Phys. Rev. B**33**, 2814 (1986);
L.A.Bendersky and M.J.Kaufman, Phil. Mag. B**53**, L75 (1986).
13 P.Ramachangrarao and G.V.S.Satry, Pramana, **25**, L225 (1985).
14 H.K.Hardy and J.M.Silcock, J. Inst. Met. **24**, 423 (1955/56);
M.D.Ball and D.J.Lloyd, Scr. Met. **19**, 1065 (1985).
15 A.-P.Tsai, A.Inoue and T.Masumoto, Jpn. J. Appl. Phys. **26**, L1505 (1987).
16 C.A.Guryan, A.I.Goldman, P.W.Stephens, K.Hiraga, A.P.Tsai, A.Inoue and T.Masumoto, Phys.Rev.Lett. **62**, 2409 (1989).

17 A.Inoue, A.-P.Tsai and T.Masumoto, to be published in J. Non-Cryst. Solids. (Proc. of 7th Int. Conf. on Liquid and Amorphous Metals, Kyoto 1989).
18 L.Bendersky, Phys. Rev. Lett. **55**, 1461 (1985).
19 K.K.Fung, C.Y.Yang, Y.Q.Zhou, J.G.Zhao, W.S.Zhan and B.G.Shen, Phys. Rev. Lett. **56**, 2060 (1986).
20 N.Wang, H.Chen and K.H.Kuo, Phys. Rev. Lett. **59**, 1010 (1987).
21 B.Dubost, J.M.Tanaka, P.Sainfort and M.Audier, Nature **324**, 48 (1986); F.W.Gayle, J. Mater. Res. **2**, 1 (1987).
22 W.Ohashi and F.Spaepen, Nature **330**, 555 (1987).
23 T.L.Ho, J.A.Jaszczak, Y.H.Li and W.F.Saam, Phys. Rev. Lett. **59** 1116 (1987).
24 K.Ingersent and P.J.Steinhardt, Phys. Rev. **B39**, 980 (1989).
25 P.A.Kalugin, A.Yu.Kitaev and L.S.Levitov, Pis'ma Zh. Eksp. Teor. Fiz. **41** 119 (1985) [Sov. Phys. - JETP Lett. **41**, 145 (1985)].
26 P.Bak, Phys. Rev. Lett. **56** 861 (1986).
27 J.E.S.Socolar, P.J.Steinhardt and D.Levine, Phy. Rev. B **32**, 5547 (1986).
28 D.Levine and P.J.Steinhardt, Phys. Rev. B **34**, 596 (1986).
29 J.E.S.Socolar and P.J.Steinhardt, Phys. Rev. B **34**, 617 (1986).
30 P.Bak, Phys. Rev. B **32**, 5764 (1985); J.E.S.Socolar, T.C.Lubensky and P.J.Steinhardt, Phys. Rev. B **34**, 3345 (1986).
31 D.Levine, T.C.Lubensky, S.Ostlund, S.Ramaswamy, P.J.Steinhardt and J.Toner, Phys. Rev. Lett. **54**, 1520 (1985).
32 P.A.Bancel, P.A.Heiney, P.W.Stephens, A.I.Goldman and P.M.Horn, Phys. Rev. Lett. **54**, 2422 (1985); V.Elser, Phys. Rev. B **32**, 4892 (1985).
33 T.C.Lubensky, J.E.S.Socolar, P.J.Steinhardt, P.A.Bancel and P.A.Heiney, Phys. Rev. Lett. **57**, 1440 (1986); P.M.Horn, W.Malzfeldt, D.P.DiVincenzo, J.Toner and R.Gambino, Phys. Rev. Lett. **57**, 1444 (1986).
34 M.V.Jaric and D.R.Nelson, Phys. Rev. B **37**, 4458 (1988); A.I.Goldman, C.A.Guryan, P.W.Stephens, J.M.Parsey,Jr., G.Aeppli, H.S.Chen and F.W.Gayle, Phys. Rev. Lett. **61**, 1962 (1988).
35 V.Elser, *Proc. of the XVth International Colloquium on Group Theoretical Methods in Physics* (Drexel University, World Sci. 1987) p.162; F.Nori, M.Ronchetti and V.Elser, Phys. Rev. Lett. **61**, 2774 (1988).
36 V.Elser and C.L.Henley, Phys. Rev. Lett. **55** 2883 (1985); Y.Ishii, Phys. Rev. B **39**, 11862 (1989).
37 M.Reed and B.Simon, *Functional Analysis* (Academic Press, New York, 1972) p.19.
38 P.A.Lee and T.V.Ramakrishnan, Rev. Mod. Phys. **57**, 287 (1985).
39 P.G.Harper, Proc. Roy. Soc. London **A68**, 874 (1955); D.R.Hofstadter, Phys. Rev. B **14**, 2239 (1976).
40 B.Simon, Adv. Appl. Math. **3**, 463 (1982).
41 J.Bellisard, R.Lima and D.Testard, Commun. Math. Phys. **88**, 207 (1983).
42 The definitions of a Liouville number and a Roth number are given in Reference 41.
43 M.Kohmoto, L.P.Kadanoff and C.Tang, Phys. Rev. Lett. **50** 1870 (1983); S.Ostlund, R.Pandit, D.Rand, H.J.Schellnhuber and E.D.Siggia, Phys. Rev. Lett. **50**, 1873 (1983).
44 A.Sütö, J. Stat. Phys. **56**, 525 (1989); S.Kotani, preprint.
45 P.A.Kalugin, A.Yu.Kitaev and L.S.Levitov, Zh. Eksp. Teor. Fiz. **91** 692 (1986) [Sov. Phys. - JETP **64**, 410 (1986)]; M.Kohmoto, B.Sutherland and C.Tang, Phys. Rev. B **35**, 1025 (1987); T.Fujiwara, M.Kohmoto.and T.Tokihiro, Phys. Rev. B **40**, 7413 (1989).

46 H.Tsunetsugu, T.Fujiwara, K.Ueda, and T.Tokihiro, J.Phys.Soc.Jpn. **55**, 1420 (1986).
47 M.Krajci and T.Fujiwara, Phys.Rev.B**38**, 12903 (1988).
48 A.R.Kortan, R.S.Becker, F.A.Thiel and H.S.Chen, Phys.Rev Lett. **64**, 200 (1989).
49 F.Spaepen, L.C.Chen, S.Ebalard and W.Ohashi, *Proc. of the Adriatico Research Conference on Quasicrystals*, edited by M.V.Jarić and S.Lundqvist (World Scientific, Singapore 1990).

Part II

Symmetry and Generalized Crystallography

Chains, Flowers, Rings and Peanuts: Graphical Geodesic Lines and Their Application to Penrose Tiling

T. Ogawa[1] *and R. Collins*[2]

[1]Institute of Applied Physics, University of Tsukuba,
Tsukuba 305, Japan and
The Institute of Statistical Mathematics, 4-6-7 Minami-Azabu,
Minato-ku, Tokyo 106, Japan
[2]Department of Physics, University of York, York YO1 5DD, UK

Abstract. The new concept of a *graphical geodesic* is introduced to characterize an arbitrary two-dimensional network whose meshes are all triangles. Although locally straight, a geodesic line globally meanders and sometimes intersects with itself. Furthermore, it may be open or closed. When the method is applied to the triangulated Penrose tiling, geodesics fall into four classes: *chains, flowers, rings and peanuts*. The analysis shows that a Penrose tiling has strong local fluctuations of curvature which average out over a small region.

1. Introduction

The Voronoï-Delaunay analysis of a set of point centres in a plane has been applied in many fields of science: the physics of disordered systems, ecology of territories, geographical problems and many others. From this point of view, local data such as number of geometrical neighbours are statistically analyzed. Some related methods for treating more global features are desirable. Such an attempt is made here. Moreover, our method is applicable for any network, not necessarily a Delaunay one.

2. Definition of Graphical Geodesics

Suppose a network whose meshes are all triangles is given. It is shown in Fig.1 with thin lines with a circle at every knot. Inscribe another triangle determined by the three midpoints of the three edges of every triangular mesh. Then paths are found all over the network which intersect each other at an edge of the mesh. The paths thus obtained are referred to as *graphical geodesics*. A geodesic line runs locally straight and globally meanders because of local curvatures in the space. Here the local curvature corresponds to the distribution of vertex coordination numbers. In view of this fact, it is useful to define the excess coordination number Δ_R of vertex R as $\Delta_R \equiv z_R - 6$ where z_R is the coordination number of vertex R.

For a regular triangular lattice where the coordination number of any vertex is six and their excess coordination numbers are all zero,

Springer Series in Solid-State Sciences, Vol. 93 Quasicrystals
Editors: T. Fujiwara · T. Ogawa © Springer-Verlag Berlin, Heidelberg 1990

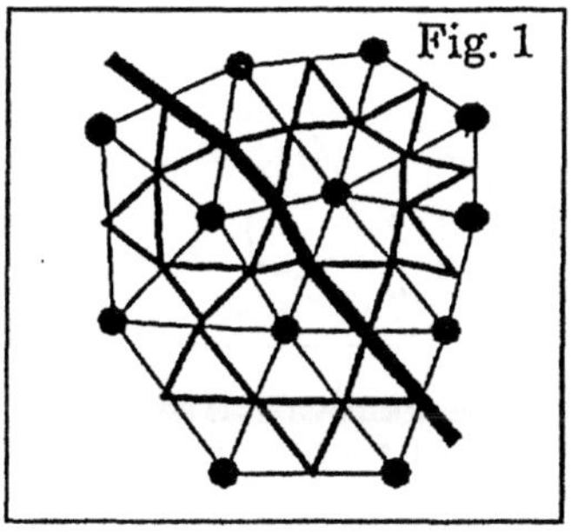

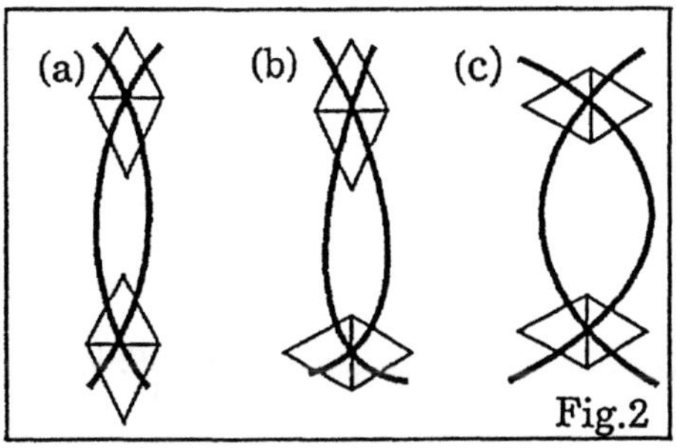

Fig. 1 A triangular network (thin lines with a circle at every knot) and its graphical geodesic lines (thick lines). A triangle determined by three midpoints of three edges of the mesh of the network is inscribed. Then the edges of the inscribed triangles form graphical geodesic lines intersecting at every edge of the meshes. A single geodesic line among them is drawn as a heavy line.

Fig.2 Three types of regions bounded by two geodesic lines. Total excess coordination numbers in the region are respectively (a) –2,(b) –3, (c) –4.

geodesic lines are three sets of parallel paths. Geodesics tend to converge (diverge) in the presence of a vertex with negative (positive) excess coordination number. If two geodesic lines never intersect, the total excess numbers of all the vertices in the infinite region between them is non-negative.

Two geodesic lines may intersect each other several times. There are three types of closed regions bounded by two geodesic lines. In the first type (Fig.2a), where the *angles* at both cross points are *acute*, the total excess coordination number in the region is –2. In the second type (Fig.2b), where the *angle* at one cross point is *acute* and that at the other is *obtuse*, the total excess coordination number in the region is –3. In the third type (Fig.2c), where the *angles* at both cross points are *obtuse*, the total excess coordination number in the region is –4.

There are three types of closed regions bounded by a single curve. In the first type (Fig.3a), where the *angle* at the cross point is *acute*, the total excess coordination number in the region is –4. In the second type (Fig.3b), where the *angle* at the cross point is *obtuse*, the total excess coordination number in the region is –5. In the third type (Fig.3c), where a geodesic line closes by itself, the total excess coordination number in the region is –6.

These can be systematically interpreted by assigning –1 and –2 respectively to the region of an *acute* and *obtuse* portion (respectively not containing and containing the edge) at a cross point (Fig.4a). Regard the region with two cross points as normal. Assign –3 to every missing cross

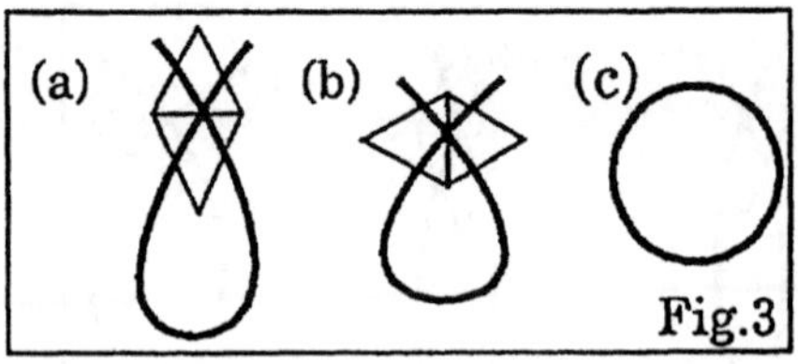

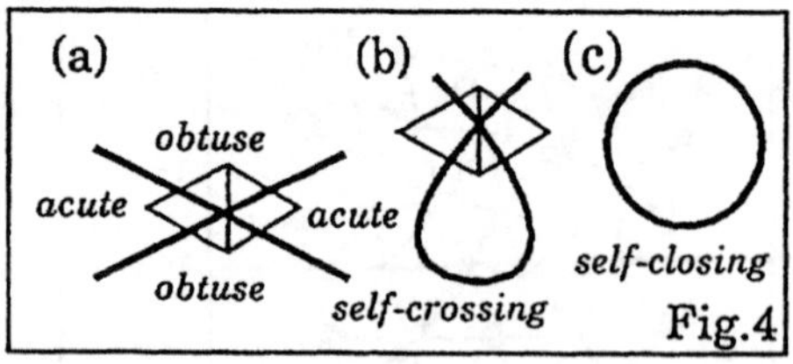

Fig.3 Three types of regions bounded by a single geodesic line. Total excess coordination number in the region are respectively (a) –4, (b) –5, (c) –6.

Fig.4 A systematic interpretation of Fig.2 and Fig.3. Regard the region with two cross points as normal. (a)Assign –1 and –2 respectively to *acute* and *obtuse* angles of a cross point. Assign –3 for each missing cross point. (b) A self-crossing case. (c) A self-closing case.

point; one is missing in a self-crossing case and two are missing in a self-closing case. Add up the two numbers thus assigned. The obtained number is equal to the total excess coordination number of the region.

It is noted that a self-closing geodesic line corresponds to a great circle in the following sense. The total excess coordination number of a triangular polyhedron is –12. The value –6 is half of that.

3. Triangulated Penrose Tiling

A triangulated Penrose network is obtained by drawing the minor diagonals of all rhombic tiles (Fig.5(a)). It is equivalent to the Delaunay network of vertex configuration of rhombic Penrose tiles. The geodesic lines of the triangulated Penrose network are classified into four classes, *chains, flowers, rings and peanuts* (Fig.5(b)–(d) and Fig.6, where the diagonals are omitted so that the location in a Penrose tiling might be easily recognized).

Chains (Fig.5(b)) are macroscopically *straight* geodesic lines. There are five sets of *parallel* geodesic lines. There are only two separation widths for two successive *parallel* geodesic lines. The ratio of these two is the golden mean. *Chains* as a whole construct the so-called Ammann lattice. A *chain* looks like a slalom avoiding S-vertices (using the name by de Bruijn). It never passes an obtuse rhombus parallel to its diagonal. The total excess coordination number is zero in any parallelogram region bounded only by two pairs of parallel chains. Exactly one third of the total length of geodesic lines are *chains* in the triangulated Penrose tiling.

Rings (Fig.5(d)) are self-closing geodesic lines of nearly circular shape. There are three sizes, of *length* 40, 60 and 80, where the *length* is measured as the number of triangles the geodesic line passes through. A *40-ring* and an *80-ring* enclose a decagon without five-fold symmetry at

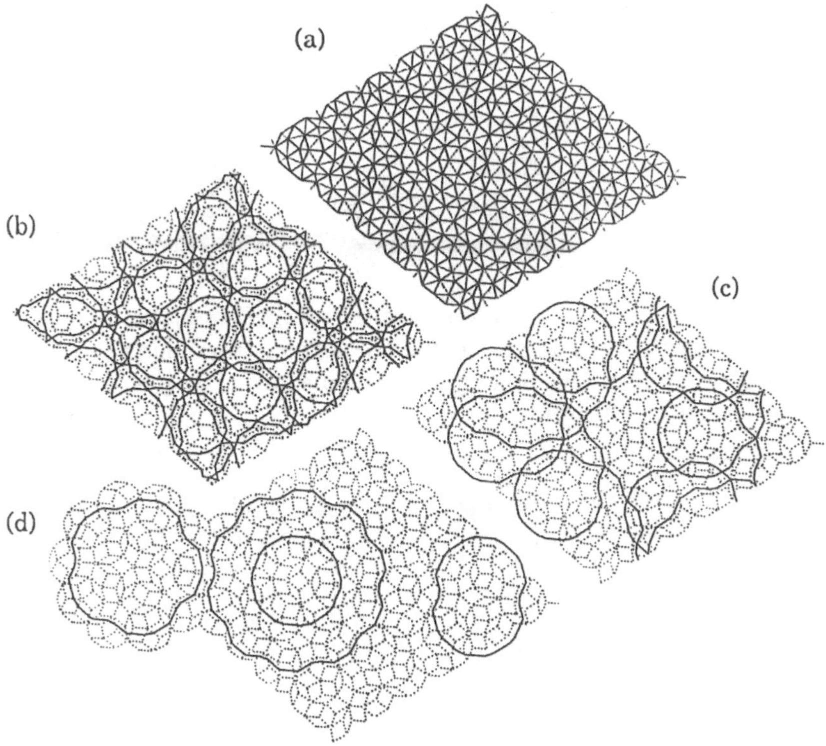

Fig. 5 (a) An example of triangulated Penrose net. All the minor diagonals are drawn in an acute rhombus of 6th generation. It exactly corresponds to the Delaunay network of vertex configuration.
(b) Ammann lattice constructed with all *chains* in the area.
(c) Parts of *flowers*.
(d) 60-, 40- (inside) and 80- (outside) *rings* and (48) *peanut* from left to right.

their centre. The centre of a *60-ring* is at a *S5* vertex and its inside has perfect five-fold symmetry. 42.75% of the total length of geodesic lines are *rings*.

The total excess coordination number inside is of course –6, corresponding to a kind of a great circle mentioned at the end of the previous section.

Peanuts (Fig.5(d)) are less symmetrical self-closing geodesic lines. There is only one kind of *peanut* of the length 48. The total excess coordination number inside is also –6. 6.50% of the total length of geodesic lines are *peanuts*.

Flowers (Fig.5(c) and Fig.6) are self-crossing and self-closing geodesic lines. They have perfect five-fold symmetry. There are two types: one,

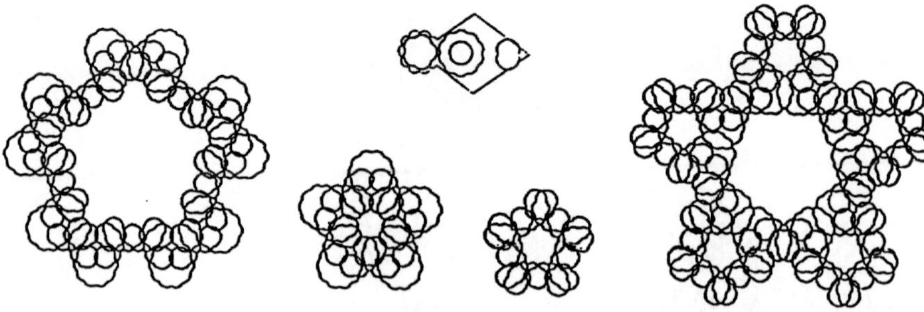

Fig. 6 3180-, 1200-, 780- and 4740-*flowers* from left to right. There are two systems of flowers: the one the former two belong to and the other the latter two belong to. A miniature of Fig.5(d) is shown as a scale of the figure.

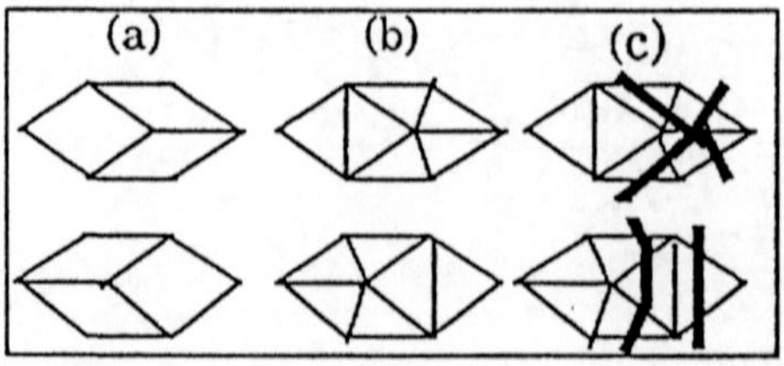

Fig.7 Recombination of geodesic lines A phason excitation corresponds to the change between the upper row and the lower. The change in the triangular network is in (b) and the recombination in (c).

with length $165 \cdot 4^n + 540 =$ 1200, 3180, 11100, 42780,..., has its centre at an S-vertex and the other, with length $330 \cdot 4^n - 540 = 780$, 4740, 20580, 83940,..., has its centre at an $S5$-vertex. Their varieties are hierarchically infinite. If coarse grained, the edges of higher order flowers are Koch curves of bottom angle 36°. 17.41% of the total length of geodesic lines are *flowers*.

It is noted that a change from a self-crossing geodesic into two self-closing geodesics takes place by rearrangement of three tiles constituting a Q-vertex, which corresponds to the excitation of a *phason* (see Fig.7). The change is a kind of bifurcation.

From the fact that there are such classes of geodesic lines, some characteristic features of Penrose tiling can be seen. There are strong fluctuations in local curvature, which result in the existence of *rings* and peanuts. These fluctuations compensate in a narrow region as can be seen from the existence of so many parallel geodesic lines. The self-similarity of Penrose tiling is reflected in the Ammann lattice of *chains* and hierarchic varieties of *flowers*. The frequency of vertices with excess coordination number 0 and ±1 is given in Table I, which actually shows rather strong fluctuations of local curvatures. For an irregular network, occurrences of so many pairs of parallel geodesic lines and so many closed geodesic lines can be expected. Existence of *flowers* must be reflected in some physical properties of Penrose tiling.

Table I Distribution of excess coordination numbers. The names of vertices are due to de Bruijn. The deviation $<\Delta^2> = 20-12\tau = 0.58$.

excess coordination number	frequency	types of vertices
-1	$10-6\tau = 29.2\%$	$Q, K, S, S5$
0	$12\tau-19 = 41.6\%$	$D, S4$
1	$10-6\tau = 29.2\%$	$J, S3$

Note that there are more fluctuations in a self-similar tiling with ten-fold symmetry with $-1 \leqq \Delta \leqq 4$ and $<\Delta^2> = 0.94$. The frequency, however, of the strong scatterers with $\Delta^2 > 1$ is rather rare. Therefore the geodesic lines go rather straight in wide flat areas and turn suddenly at a strong scatterer. This situation is also seen in the self-similar tiling of octagonal symmetry with squares and 45°-135°-rhombi ($-1 \leqq \Delta \leqq 2$ and $<\Delta^2> = 1.03$). In these cases, there is a *ring* geodesic at every strong scatterer since the excess coordination number at the n-fold centre is $n-6$ and that of the n neighbours is -1 (when $n \geqq 8$): $(n-6) + n \times (-1) = -6$. It is noted that the area of averaging out is wide in these tilings with a large deflation ratio. One cannot help noticing how high the uniformity of the original Penrose tiling is.

The Ehrenfest Process, a Rotational Representation of Spin Operators and Explicit Forms of Rotation Operators

T. Ogawa

Institute of Applied Physics, University of Tsukuba,
Tsukuba 305, Japan and
The Institute of Statistical Mathematics, 4-6-7 Minami-Azabu,
Minato-ku, Tokyo 106, Japan

Abstract. One may think the topics vary too much: the Ehrenfest process (a stochastic model), spin operators and rotation operators. These are unified in the context of completion of icosahedral harmonics described in Sect.1. The problems are not only technical but more fundamental. Apart from the context, the connection between the Ehrenfest process and the spin operator S_x (not Ising spin) seems to suggest some hidden relations between classical physics and quantum physics.

1. Introduction

Icosahedral symmetry can be considered as being beyond three-dimensional. In treating cubic symmetry, a knowledge of group theory is not always necessary so long as one utilizes the symmetry in variables x, y and z and their signs: there are $48=3!\cdot 2^3$ equivalent points altogether where 3! is the number of permutations of x, y and z and 2^3 is that of combinations of three signs $\pm$. However, in treating icosahedral symmetry in which there are 120 equivalent points in general, it is very difficult because one must regard the points with different appearance as equivalent. Something about icosahedral symmetry is out of our three-dimensional intuition. An attempt is made here to explore some tools to treat icosahedral symmetry.

Cubic harmonics have been investigated for a long time for application to the study of the effects of the crystalline field. Spherical harmonics, being the base of the rotation group, are resolved into the irreducible representation of cubic group l by l, which specifies irreducible representation of the rotation group. For example, the 5-dimensional representation $l=2$ is reduced into 2-dimensional E_g and 3-dimensional T_{2g} as is well known. The reduction table of the l-subspace, which is the irreducible representation of the rotation group, by the icosahedral group is obtained in Sect. 5 by making use of the character table. If k-fold rotation is an element of the concerned group, the smallest common multiple of all k's is the period of the cyclic part of the table. In the cubic case with 2-, 3- and 4-fold rotations, for example, it is 12. It is 30 in the icosahedral case with 2-, 3- and 5-fold rotations.

Springer Series in Solid-State Sciences, Vol. 93 Quasicrystals
Editors: T. Fujiwara · T. Ogawa

What is much more tedious is to obtain their explicit functional forms. The harmonics are polynomials of l-th power and it is necessary to prepare some systematic tool in order to treat up to $l=29$. In the matrix representation, a $(2l+1)\times(2l+1)$ matrix must be managed. The motivation of the following attempt is as mentioned above.

In spite of the context just mentioned above, the starting point and the main part of this work are rather different. It concerns firstly a new useful representation of spin or angular momentum operators and secondly a systematic method of obtaining an explicit form of any rotational operator for any l.

The transition matrix of the Ehrenfest process is studied in Sect. 2. It is shown that the results in Sect. 2 can be used as a representation of spin operators or angular momentum operators in Sect. 3. It can also be applied to get the explicit form of any rotation matrix in Sect. 4. The reduction table of irreducible representations from the rotational group to the icosahedral one is given in Sect. 5. Some lowest order icosahedral harmonics are shown in new expressions in Sect. 6.

2. Transition Matrix of the Ehrenfest Process

The Ehrenfest process is a stochastic process proposed by P. and T. Ehrenfest [1] in 1907 in connection with reconciliation of time reversibility and recurrence with observable irreversible behavior. Suppose N balls numbered from 1 to N are distributed in two boxes A and B. An integer from 1 to N is chosen at random and the ball corresponding to that number is moved from the box in which it is to the other one. The process is then repeated a desired number of times.

So long as only the number of balls in box A is of interest, the $(N+1)\times(N+1)$ transition matrix is given by

$$[W_N]_{k'k} = \frac{1}{N}\left\{(N-k)\delta_{k'\,k+1} + k\delta_{k'\,k-1}\right\} \qquad (k',k=N,N-1,\ldots,0) \tag{2.1}$$

where (k',k)-element $[W_N]_{k',k}$ stands for the conditional probability that a change to k' occurs when k balls are in A. The eigenvalues of the matrix W_N are arranged periodically with 1 and -1 as the two ends; [2]

$$\lambda_k = \frac{2k}{N} - 1, \qquad (k=N,N-1,\ldots,0). \tag{2.2}$$

This fact reminds a physicist of a spin operator. In the next section, a rational representation of spin and/or angular momentum operators is presented. Let us investigate the eigenvectors of W's before that. The eigenvector associated with eigenvalue λ can be expressed in the form of a generating function

$$F_{N,\lambda}(t) = \frac{1}{2^{N/2}}(t+1)^{(1+\lambda)N/2}(t-1)^{(1-\lambda)N/2} = \sum_{k=0}^{N} V_k(\lambda)t^k, \tag{2.3}$$

where the normalizations are chosen so that

$$V_N(\lambda)=(-1)^{N(\lambda-1)/2}V_0(\lambda)=2^{-N/2}. \tag{2.4}$$

Matrix V is defined by arranging these eigenvectors as its column in descending order of eigenvalues $\lambda=(2k'/N)-1$ as

$$V_{kk'}=V_k(\frac{2k'}{N}-1) \tag{2.5}$$

with a remarkable property

$$V^2=E, \quad V^{-1}=V, \tag{2.6}$$

where E is the $(N+1)\times(N+1)$ unit matrix. The relation is partly attributed to the normalization factor $2^{-N/2}$ in (2.3).

3. Rational Representation of Spin Operators

Matrix W_N in the previous section can be transformed into the x component of spin of magnitude $S=N/2$

$$SW_N=T^{-1}S_xT, \quad STW_NT^{-1}=S_x \tag{3.1}$$

where the transformation matrix T is given by

$$T_{kk'}=\frac{1}{\sqrt{{}_NC_k}}\delta_{kk'}, \quad [T^{-1}]_{kk'}=\sqrt{{}_NC_k}\,\delta_{kk'}. \tag{3.2}$$

It is a kind of affine transformation. At the same time, matrix V in the previous section is transformed into U as

$$TVT^{-1}=U, \quad V=T^{-1}UT, \tag{3.3}$$

where the k-th column of matrix U is the eigenvector of S_x with eigenvalue $k-S$ ($k=2S,2S-1,\dots 2,1,0$).

It is noted that SW_{2S} can be regarded as a representation of S_x where all the matrix elements are rational numbers. This property transfers to the eigenvectors and then matrix V. It is remarkable that all the elements of V are expressed in the form of a generating function, however large the magnitude of the spin. This property can be extended for any direction of spin other than x. The generating function for the eigenvector of spin $S_r=(\mathbf{S}\cdot\mathbf{r})$, where $\mathbf{S}=(S_x, S_y, S_z)$ of direction $\mathbf{r}=(a, b, c)=(\sin\theta\cos\phi, \sin\theta\sin\phi, \cos\theta)$ is given by

$$F^{(a,b,c)}_{N,\lambda}(t)=\left(\frac{-1}{2t}\right)^S\left((1+c)t+(a+bi)\right)^{S+m}\left((a-bi)t-(1+c)\right)^{S-m}, \tag{3.4}$$

$$F^{(\theta,\phi)}_{N,\lambda}(t)=\left(\frac{-1}{2t^*}\right)^S\left(t^*\cos\frac{\theta}{2}+\sin\frac{\theta}{2}\right)^{S+m}\left(t^*\sin\frac{\theta}{2}-\cos\frac{\theta}{2}\right)^{S-m}, \tag{3.5}$$

$$t^*\equiv t\exp(-i\phi)$$

where the expansion parameter is always t (not t^*). Thus, one can obtain the eigenvectors and matrix $V(\theta, \phi)$ for any direction of spin with any magnitude by using these generating functions. Any $V(\theta, \phi)$ has the same property as (2.5).

4. Rotation Operators

4.1 180°-rotation.

The property of V shown in (2.5) suggests that V may be a transformation matrix of 180°-rotation around some axis. Actually, U, which is the transformation of V by T as shown in (3.3), is a transformation matrix of 180°-rotation around the bisector of the x-axis and z-axis as

$$U_S S_x U_S^{-1} = S_z, \quad U_S S_y U_S^{-1} = -S_y, \quad U_S S_z U_S^{-1} = S_x. \tag{4.1}$$

Similarly, the $V(\theta, \phi)$ corresponding to any $U(\theta, \phi)$ is a transformation matrix of 180°-rotation around the bisector of the (θ, ϕ)-*direction* and z-axis in the rational representation.

In conclusion, one can use rational representation in an actual calculation and finally transform the obtained results into the conventional form by T.

4.2 Rotation of any angle

A rotation of angle θ around a unit vector **O** can be decomposed into two 180°-rotations; first around a unit vector **A** and then around another unit vector **B**: The two unit vectors **A** and **B** are both on a great circle perpendicular to **O**, making an angle $\theta/2$ as

$$(\mathrm{OA}) = (\mathrm{OB}) = 0 \quad \text{and} \quad (\mathrm{AB}) = \cos\frac{\theta}{2}. \tag{4.2}$$

More precisely, **B** is decided by an arbitrarily chosen **A** on the great circle of axis **O** by the rotation of angle $\theta/2$ around **O** as

$$\mathrm{B} = \mathrm{A}\cos\frac{\theta}{2} + [\mathrm{O} \times \mathrm{A}]\sin\frac{\theta}{2}. \tag{4.3}$$

4.3 Rotation operator

It is well known that the operator expressing a rotation of angle θ around unit vector **O** is given by

$$R(\mathrm{O}, \theta) = \exp[-i\theta(\mathrm{O}\cdot\mathrm{S})], \tag{4.4}$$

where $\mathrm{S} = (S_x, S_y, S_z)$ is a spin operator vector and its magnitude is arbitrary. Knowing the definition, one cannot obtain its explicit matrix expression easily since it is necessary to diagonalize the operator in the exponent, which is a rather tedious job especially for large spin magnitudes although it is possible in principle. Then the rational representation developed in this paper is useful to get explicit forms of rotation operators for any magnitude of S. By making use of (3.4) or (3.5), the explicit form of $R(\mathrm{O}, 180°)$ is easily found and that for arbitrary θ is found by following the discussion in Sect. 4.2.

5, Reduction of l-Subspace by Icosahedral Symmetry

The scheme of how l-subspace, which is an irreducible representation of rotation group, is reduced by icosahedral symmetry is listed in Table I up to $l=15$.

It is noted that the corresponding table for cubic symmetry is up to $l=6$ as pointed out in 1. The discussion up to the previous section is not necessary for obtaining Table I; the character table alone is enough. However, the table says that a polynomial of 29th power or 59×59-matrix must be treated to get icosahedral harmonics in the first period of the cycle $0 \leqq l \leqq 29$. This is why the discussions up to Sect. 4 are useful for this purpose.

Table I Reduction of l-subspace by icosahedral symmetry

The table lists the coefficient n_k of the expansion $2l+1=\Sigma n_k \Gamma_k$ where Γ_k is the k-th irreducible representation of the I_h group.For example, the 7-dimensional subspace $l=3$ is reduced to 3-dimensional T_{2u} and 4-dimensional G_u. It is noted that addition of the corresponding numbers, (0, 1, 1, 2, 3) for $l=14$ and (1, 2, 2, 2, 2) for $l=15$, gives the dimensionality (1, 3, 3, 4, 5) of every irreducible representation. This relation holds for any pair l_1 and $l_2=29-l_1$. Add the dimensionality (1, 3, 3, 4, 5) to the numbers for $l'=l-30$ for $30 \leqq l \leqq 59$, Thus the list is cyclic in l in a sense.

l (even)						l (odd)					
l	A_g	T_{1g}	T_{2g}	G_g	H_g	l	A_u	T_{1u}	T_{2u}	G_u	H_u
0	1	0	0	0	0	1	0	1	0	0	0
2	0	0	0	0	1	3	0	0	1	1	0
4	0	0	0	1	1	5	0	1	1	0	1
6	1	1	0	1	1	7	0	1	1	1	1
8	0	0	1	1	2	9	0	1	1	2	1
10	1	1	1	1	2	11	0	2	1	1	2
12	1	1	1	2	2	13	0	1	2	2	2
14	0	1	1	2	3	15	1	2	2	2	2

6. Some Examples of Icosahedral Harmonics

Icosahedral harmonics up to $l=3$ are listed in Table II. The expressions adopted are as symmetrical as possible. Those of $l=0$ and 1 are the same as usual. In the expression for $l=2$, all five are equivalent though some influence of either the monoaxial or cubic viewpoint remains in the conventional expression. It is desirable to associate five 3D-vectors to the basis of the 5-dimensional representation, if possible. Five directions, however, cannot be chosen evenly in three dimensions. Then the alternative way is to take six 5-fold axes e_i (i=0,1,2,3,4,5) of icosahedral symmetry

$$\begin{aligned} (\mathrm{e}_0 \mathrm{e}_i) &= (\mathrm{e}_i \mathrm{e}_{i+1}) = 1/\sqrt{5}\,, \quad && (i=1,2,3,4,5) \\ (\mathrm{e}_i \mathrm{e}_{i+2}) &= -1/\sqrt{5}\,, && (i=1,2,3,4,5) \qquad (6.1) \\ &\text{where} && \mathrm{e}_6=\mathrm{e}_1 \text{ and } \mathrm{e}_7=\mathrm{e}_2 \end{aligned}$$

as an overcomplete set and to orthogonalize five of them properly afterwards by making use of the other one as the common element so that all five are equivalent. Here, the variable is the unit vector r.

In the expression for G_u in $l=3$, which is another special feature of the icosahedral group, it is natural to associate four tetrahedrally chosen vectors with the basis. It is also natural to choose the four vectors from among 20 three-fold axes. The choice contains a sort of chirality problem. The fact is reflected in the expression.

The vectorial representation of spherical harmonics, part of which is used here for H_g, is more natural than the Cartesian one in the sense that expressions are closer to spherical symmetry. Its more systematic proposition will be published elsewhere.

Table II Icosahedral harmonics up to $l=3$

$l=0$	A_g	1
$l=1$	T_{1u}	$x,\ y,\ z$
$l=2$	H_g	$\Psi_i(\mathrm{r}) = \frac{1}{2\sqrt{30}}\left[(1+\sqrt{6})[3(\mathrm{e}_0\mathrm{r})^2 - \sqrt{6}\mathrm{r}^2] + 15(\mathrm{e}_i\mathrm{r})^2\right]$ $(i=1,2,3,4,5)$
$l=3$	T_{2u}	$\frac{\sqrt{3}}{6}x\left[x^2+\frac{3}{\tau}y^2-3\tau z^2\right]$, $\frac{\sqrt{3}}{6}y\left[y^2+\frac{3}{\tau}z^2-3\tau x^2\right]$, $\frac{\sqrt{3}}{6}z\left[z^2+\frac{3}{\tau}x^2-3\tau y^2\right]$
	G_u	$\frac{1}{4}\left[(x^3+y^3+z^3)-\tau^2(xy^2+yz^2+zx^2)-\tau^{-2}(xz^2+yx^2+zy^2)+2\sqrt{15}xyz\right]$
		$\frac{1}{4}\left[(x^3-y^3-z^3)-\tau^2(xy^2-yz^2-zx^2)-\tau^{-2}(xz^2-yx^2-zy^2)+2\sqrt{15}xyz\right]$
		$\frac{1}{4}\left[(-x^3+y^3-z^3)-\tau^2(-xy^2+yz^2-zx^2)-\tau^{-2}(-xz^2+yx^2-zy^2)+2\sqrt{15}xyz\right]$
		$\frac{1}{4}\left[(-x^3-y^3+z^3)-\tau^2(-xy^2-yz^2+zx^2)-\tau^{-2}(-xz^2-yx^2+zy^2)+2\sqrt{15}xyz\right]$

7. Concluding Remarks

A relation between the transition matrix of the Ehrenfest process and the spin operator S_x is pointed out in Sects. 2-4. Though the relation with Ising spin is trivial, the relation here is with quantum or dynamic spin. The elements of a vector for matrix W stand for probabilities if they are all positive. Their squares stand for probabilities in spin S_x. Though the author does not know what lies in their equivalence, something fundamental may be concerned. The problem seems not only technical.

References

1. P. Ehrenfest and U. Ehrenfest, Phys. Z. 8, 311 (1907)
   P. Ehrenfest and U. Ehrenfest, Encyklop. d. math. Wissensch. IV 2, II (1911).
2. M. Kac: *Probability and Related topics in Physical Sciences* (1957, Interscience Publishers, London)

Generalized Crystallography in Two and Three Dimensions

N.D. Mermin

Laboratory of Atomic and Solid State Physics, Cornell University,
Ithaca, NY 14853, USA

Abstract. We have reformulated crystallography in two and three dimensions to include quasicrystallographic structures with point group symmetries forbidden by conventional crystallography. The trick is to develop the subject entirely in wave-vector space, where a Bragg-like diffraction pattern always defines a lattice of wave-vectors. This generalized reciprocal lattice need not have a shortest distance between its points, so its point group symmetry need not be crystallographic, nor need the lattice be dual to a direct lattice of real space translations. One first classifies the reciprocal lattices with a given point group symmetry. This turns out to be a non-trivial exercise, even in two dimensions. Then, for a given point group and lattice, one classifies the possible phase relations of density Fourier coefficients at symmetry related points, which gives the generalized space groups. In the quasicrystallographic case lattice scale invariance can simplify the space-group classification, leading to an identification of space-groups that would otherwise be viewed as distinct. Because quasicrystallographic reciprocal lattices have no real space direct lattices, tiling theory is a necessary part of quasicrystallography: aperiodic tilings provide simple templates for arranging atomic units into structures with density Fourier coefficients characterized by the various space groups. Generalizations of conventional tiling algorithms are needed to construct such tilings.

In 1962 Bienenstock and Ewald [1] pointed out that the 230 three-dimensional crystallographic space-groups could be enumerated more systematically and efficiently in reciprocal space. David Wright, Dan Rokhsar, and I [2-5] maintain that this approach is the key to understanding the space-group classification of three dimensional quasicrystals; indeed, we would argue that ordinary three-dimensional crystallography is best understood as a subset of the more compact and coherent scheme of three-dimensional quasicrystallography.

The alternative to our approach is to view quasicrystallographic structures as lower dimensional projections of higher dimensional periodic structures [6-12], extracting their space-groups from the higher dimensional crystallographic ones. But this requires a prior knowledge of crystallography in arbitrarily high dimensions even to yield such simple results as those summarized in the table below. It also obscures the pleasing view of the d-dimensional crystallographic lattices and space groups as simply the bottoms of towers of their d-dimensional quasicrystallographic descendants (a program, to be sure, that we have currently completed only in the two-dimensional case).

Reciprocal space is the natural place to construct a crystallography broad enough to include quasicrystals, because the hallmark of a quasicrystal is a diffraction pattern with sharp Bragg peaks and a non-crystallographic point group symmetry. The density has non-zero Fourier coefficients at wave-vectors in the diffraction pattern. Because the density is determined in a highly non-linear way (consider, for example, Ginzburg-Landau theories), if the density has Fourier coefficients at two vectors it will in general also have them at the sum and difference vectors. The wave-vectors with non-vanishing density Fourier components thus lie inside a set that is closed under addition and subtraction — the defining property of a lattice [13]. This is the Bravais lattice, the skeleton from which the flesh of crystallography hangs. It is the set of all integral linear combinations of wave-vectors determined by the peaks in the diffraction pattern.

Of course any real diffraction pattern has only a finite number of peaks, because of unavoidable limits on instrumental resolution. The lattice of wave-vectors is established in practice by observing a relatively small number of peaks; adding in wave-vectors from subsequent measurements does not further expand the lattice.

In the crystallographic case the lattice of wave-vectors determined by the diffraction pattern is the ordinary reciprocal lattice, dual to a lattice of real-space translations under which the periodic real-space density is invariant. In the quasicrystallographic case the lattice is not dual to a lattice of real-space translations: you are stuck in wave-vector space. The other striking difference is that there is a minimum distance between wave-vectors in the lattice determined by the crystallographic diffraction pattern, but the lattice determined by a quasicrystallographic diffraction pattern is dense in k-space: there is no minimum distance between its points. Since observed diffraction patterns never contain more than a finite subset of the lattice, the manifestation of no minimum distance between lattice points is that more peaks can always be resolved between already established peaks as the resolution gets better.

Lacking a minimum distance between points one cannot prove the famous theorem that a lattice in two or three dimensions can only have 2-, 3-, 4-, or 6-fold symmetry axes, so the symmetry groups of quasicrystallographic diffraction patterns are not restricted to the crystallographic point groups. Without a minimum distance between points, however, there are too many candidates for quasicrystallographic lattices. Indeed, objects are admitted that do not qualify as reciprocal lattices even when the point group is crystallographic. Consider, for example, the set of all integral linear combinations of two orthogonal unit vectors and another pair that are the first two scaled up by $\sqrt{2}$. This is a lattice (closed under addition and subtraction) which clearly has square symmetry and is not equivalent to the ordinary square lattice. It is not counted as a crystallographic lattice because it is clearly a superposition of two incommensurate square lattices, and is better regarded not as a fundamental crystallographic object, but an incommensurately modulated structure. What singles out the crystallographic lattices in d-dimensions is that all their points are integral linear combinations of just d vectors. For quasicrystallographic lattices in d dimensions, d vectors are not enough, but for any given non-crystallographic point group symmetry there is a minimum number $D > d$ of vectors, the integral linear combinations of which can constitute a lattice with that point group symmetry. We call D the indexing dimension of the point group, and regard as quasicrystallographic lattices only those that are integral linear combinations of a number of vectors (the "rank" of the lattice) equal to the indexing dimension of their point group. Like crystallographic lattices, quasicrystallographic lattices have the minimum rank compatible with their point group symmetry. Lattices with a rank bigger than symmetry requires, should be viewed as superpositions.

With this definition in hand one can embark upon the first part of the quasicrystallographic program: enumerating the lattices with an arbitrary point group. The result for icosahedral lattices [2,6] confirms what many people had earlier guessed: there are three distinct lattices given by integral linear combinations of six vectors of equal length along the six distinct 5-fold axes with (1) all integral coordinates (P^*), (2) integral coordinates constrained to have an even sum (F^*), and (3) integral coordinates constrained to be all odd or all even (I^*). Establishing this result requires a simple but non-trivial number theoretic theorem. Define the "golden integers" to be the one-dimensional lattice given by all integral linear combinations of unity and the golden mean, $\tau = (\sqrt{5}+1)/2$ and consider any subset of the golden integers closed under addition and subtraction. The theorem states that any such sublattice of the golden integers that is invariant under a rescaling by τ is nothing but a rescaled version of the golden integers themselves. This theorem — in mathematical terminology the fact that all ideals in the ring of golden integers are principal ideals — and its generalizations are of crucial importance in investigating the numbers of quasicrystallographic lattices.

Aside from the icosahedral lattices, the three-dimensional quasicrystallographic lattices all have axial point groups, and the condition of minimum rank restricts them to uniform stackings of two-dimensional sublattices a finite distance apart. Two-dimensional lattices have even rotational symmetry, since they include the negatives of all their vectors (which in two dimensions are given by rotations through π). The simplest are given by all integral linear combinations of N unit vectors spaced $2\pi/N$ apart. These "standard lattices" are most efficiently viewed in the complex plane, as the set of all integral linear combinations of the N-th roots of unity. In this guise they are known to mathematicians as cyclotomic integers of degree N. (The mathematical convention is to take the degree to be $N/2$ when N is twice an odd number.)

The question immediately arises whether there can be non-standard two-dimensional lattices, or whether all are simply rotated and rescaled versions of the standard ones [3]. (When $N = 2$, equivalent lattices can be related by more than just rotations and rescalings, but this special case is crystallographic and, of course, well understood.) The corresponding mathematical question is whether all ideals of the cyclotomic integers are principal ideals, and the surprising answer is that in general they are not. However all two-dimensional lattices are equivalent to the standard lattice for all N from 4 to 44 (and for $N = 48, 50, 54, 60, 66, 70, 84$, and 90.) For all other N there are non-standard lattices, and the number of distinct lattices can be enormous (for example 359,057 when N=128).

Until quasicrystals are discovered with rotational symmetry axes bigger than 22-fold, however, it is safe to restrict one's attention to the standard lattices. The obvious way to stack standard lattices into a 3-dimensional axial lattice is by placing the layers directly above one another, just as square lattices stack to give simple tetragonal, or triangular lattices stack to give hexagonal. The layers can also stack with a uniform shift from layer to layer, as triangular is stacked to give trigonal or square to give centered tetragonal. In enumerating the possibilities we impose the requirement that the N-fold rotational symmetry of the two-dimensional sublattices should be no greater than required by the n-fold symmetry of the three-dimensional lattice in which they are embedded. This means $N = n$ for even n and $N = 2n$ when n is odd. The case of odd n can only arise when the stacking is staggered, since a vertical stacking gives $N = n$ and N is necessarily even. It is easy to show that if n has two distinct prime factors the stacking must be vertical, and therefore one can have axial lattices with n-fold symmetry for odd n only if n is a power of a single prime number, $n = p^j$. A less trivial result, which depends crucially on the scale invariances of the standard quasicrystallographic lattices, is that when $n = p^j$ all staggered stackings are equivalent (to within a rotation and independent horizontal and vertical rescalings) to a single staggered one that repeats every p layers [14]. When n is even the only stackings are vertical, as in the crystallographic hexagonal case, except when n is a power of 2, in which case, following the rule for the odd prime powers, there is again a single staggered stacking that repeats every two layers [14].

Having defined and computed the quasicrystallographic lattices, one next defines and computes the quasicrystallographic space groups. A material with a point diffraction pattern is said to have a point group G if its density Fourier coefficients satisfy

$$\rho(g\mathbf{K}_1)\cdots\rho(g\mathbf{K}_j) = \rho(\mathbf{K}_1)\cdots\rho(\mathbf{K}_j) \tag{1}$$

for all operations g in G and for all sets of wave-vectors in the diffraction pattern that sum to zero. This is just the condition that all macroscopic translationally invariant properties of the material should have the symmetry of the point group G.

It follows from condition (1) that for each element g of G there must be "phase functions" $\Phi_g(\mathbf{K})$ that are linear (modulo 1) on the lattice and relate density Fourier coefficients at symmetry related wave-vectors:

$$\rho(g\mathbf{K}) = \exp\bigl(2\pi i\Phi_g(\mathbf{K})\bigr)\,\rho(\mathbf{K}). \tag{2}$$

(When there are extinctions (see below) then the lattice is bigger than the diffraction pattern, and it might appear from (2) that the phase functions are only defined on the diffraction pattern, but it is easy to show that they have a unique linear extension to the entire lattice.) The space-group classification of quasicrystals is an organization into classes of equivalent phase functions according to two criteria:

Gauge Equivalence. Two densities have identical macroscopic translationally invariant properties if their Fourier coefficients are related by

$$\rho'(\mathbf{K}) = \exp\left(2\pi i\chi(\mathbf{K})\right)\rho(\mathbf{K}) \tag{3}$$

where the "gauge function" χ is linear (modulo 1) on the lattice. The corresponding phase functions are related by

$$\Phi'_g(\mathbf{K}) \equiv \Phi_g(\mathbf{K}) + \chi(g\mathbf{K} - \mathbf{K}), \tag{4}$$

where "$\equiv$" denotes equality modulo 1. Two sets of phase functions related by (4) for every element g in G are called "gauge equivalent". In the crystallographic case gauge equivalent phase functions describe structures that differ only by a real space translation. In the quasicrystallographic case gauge equivalent densities can in addition differ by a phason.

Scale Equivalence. If there is a continuous family of symmetry preserving deformations T_λ taking one lattice into another, the two are equivalent. The Fourier coefficients $\rho(\mathbf{K})$ and $\rho(T_\lambda\mathbf{K})$ describe different densities with the same space groups on different but equivalent lattices. If, however, a particular member of the family, T_{λ_0}, leaves the lattice unchanged, then $\rho(\mathbf{K})$ and $\rho(T_{\lambda_0}\mathbf{K})$ will describe different densities, still with the same spacegroup, but on the same lattice. Thus

$$\Phi'_g(\mathbf{K}) \equiv \Phi_g(T_{\lambda_0}\mathbf{K}) \tag{5}$$

will characterize the same space group class as $\Phi_g(\mathbf{K})$. Unlike gauge equivalence, scale equivalence is peculiar to quasicrystals: crystallographic lattices have no non-trivial scale invariance.

For a given lattice and point group the computation of the classes of phase functions — *i.e.* the spacegroups — is based on the following considerations [1]: (*a*) because they are linear (modulo 1) it suffices to know the phase functions for the vectors $\mathbf{b}_i$, $i = 1 \dots D$ that generate the lattice; (*b*) the group condition $\rho(g(h\mathbf{K})) = \rho((gh)\mathbf{K})$ leads to the condition

$$\Phi_{gh}(\mathbf{K}) \equiv \Phi_g(h\mathbf{K}) + \Phi_h(\mathbf{K}) \tag{6}$$

so it is only necessary to know the phase functions for a set of elements that generate the point group G; (*c*) the structure of the point group G is entirely determined by the set of relations obeyed by its generators.

Thus for the proper icosahedral group Y there are two generators t and u that satisfy the relations:

$$t^5 = u^2 = (ut)^3 = 1. \tag{7}$$

These generating relations lead to the corresponding conditions on the phase functions

$$\Phi_{t^5} \equiv \Phi_{u^2} \equiv \Phi_{(ut)^3} \equiv 0. \tag{8}$$

When expanded using properties (*a*) and (*b*) the generating relations (8) lead to a set of linear congruences for $\Phi_t(\mathbf{b}_i)$ and $\Phi_u(\mathbf{b}_i)$. The distinct classes of solutions to such congruences give the space groups.

The conventional terminology of crystallographic space groups can be applied to the quasicrystallographic case provided that terminology is defined in terms of phase functions rather than real space symmetries. Thus a *symmorphic* space group is one for which there is a gauge in which $\Phi_g(\mathbf{K}) \equiv 0$ for all g and $\mathbf{K}$. Since taking all phase functions to be zero satisfies all the conditions, there is always a symmorphic space group for any point group and lattice. Non-symmorphic space groups are said to contain a *screw axis* if there is no gauge in which Φ_r is equivalent to 0 for some rotation r, and a *glide plane* if Φ_m cannot be equivalent to 0 for some mirror m.

Non-symmorphic point groups can lead to extinctions. If $\Phi_g(\mathbf{K}) \not\equiv 0$ and $g\mathbf{K} = \mathbf{K}$, a pair of conditions that (4) ensures are gauge invariant, then

$$\rho(\mathbf{K}) = \rho(g\mathbf{K}) = \exp\bigl(2\pi i \Phi_g(\mathbf{K})\bigr)\rho(\mathbf{K}), \tag{9}$$

and this requires $\rho(\mathbf{K})$ to vanish.

From our point of view a density characterized by a space group with extinctions should not be regarded as having peaks missing from its diffraction pattern, but as having a diffraction pattern which symmetry requires to be smaller than the lattice it gives rise to.

The result of applying this procedure to a determination of the icosahedral space groups is surprisingly simple. Since there are two possible point groups, the proper icosahedral group Y or the full icosahedral group Y_h, and there are three possible lattices, there are at least the six symmorphic space groups given by making all phase functions zero. In addition, when the point group is Y there is a single non-symmorphic space group for each of the three reciprocal lattices; when the point group is Y_h there is a single non-symmorphic space group for the P^* and I^* lattices, but none for the F^* lattice. There are thus a total of eleven distinct icosahedral space groups.

Scale equivalence is again crucial in establishing these results [5,6]. If only gauge equivalence is taken into account, when the point group is Y one finds four gauge inequivalent 5-fold screw axes for each of the three lattices, but all four types are scale equivalent. When the point group is Y_h the lattice I^* has three gauge inequivalent glide planes but all three are scale equivalent.

A third example of the importance of lattice scale invariance arises in classifying two dimensional space groups [4]. The two-dimensional triangular lattice has a pair of crystallographic space groups, $p3m1$ and $p31m$, that differ only in the orientation of the point group with respect to the lattice: in $p3m1$ the mirrors lie along the lines that join the origin of the reciprocal lattice to its nearest neighbors; in $p31m$ the mirrors are perpendicular to those lines. This possibility of orienting n mirror lines in two distinct ways can arise only when n is odd, $n = 3$ being the only crystallographic example.

In the quasicrystallographic case all odd numbers n are available. But the lattice is dense and one cannot identify nearest or next nearest neighbors of the origin. One can, however, identify stars of n vectors whose integral linear combinations generate the lattice. In the case of the standard lattice it turns out that such stars have a unique set of orientations if and only if n is a power of a single prime number p. Thus if $n = p^j$ one can distinguish between $pnm1$ and $pn1m$ depending on whether the mirrors lie along or perpendicular to the directions of the

2D Space Groups With Standard Lattices (n > 2)

Point group order, n	*Lattice order, N*	*Space groups*	*Symmorphic?*	*Crystallographic*
$2^j, j > 1$	$N = n$	*pn*, *pnmm* *pngm*	yes no	4
other even n	$N = n$	*pn*, *pnmm*	yes	6
p^j, p odd prime	$N = 2n$	*pn*, *pnm*1, *pn*1*m*	yes	3
other odd n	$N = 2n$	*pn*, *pnm*	yes	none

vectors in the stars. If n is not a power of a single prime, however, one can show that generating stars lie along two distinct sets of n directions, the directions in one set being the normals to the directions in the other. There is then no way to distinguish between *pnm*1 and *pn*1*m* so the standard axial lattice with point group *nm* has only a single space group *pnm*.

We have computed all two-dimensional quasicrystallographic space groups when the lattice is standard [4] (and thus our tabulation is complete when the rotational order of the space group is less than 23). The results of the computation are given in the table below. There are two noteworthy features. One is the abovementioned collapse of *pnm*1 and *pn*1*m* into a single *pnm* when n is odd but not a prime power. The other is that non-symmorphic two dimensional space groups can arise only when n is a power of 2.

The tabulation does not include the cases n=1 or 2, which as noted above, are more complicated, but also crystallographic and therefore well known. From this broader perspective, the traditional catalog of seventeen two-dimensional space groups owes its complexity to the pathologies of $n = 1$ and $n = 2$. The 3-, 4-, and 6-fold crystallographic cases fit neatly into the paradigm of the table.

David Rabson, Jason Ho, and I have produced real space quasicrystallographic examples of these two-dimensional space groups, by generalizing the familiar tiling algorithm. To make a conventional Penrose tiling you divide a plane — the "grid plane" — into cells by superposing n grids of uniformly spaced parallel lines whose normals differ in direction by $2\pi/n$. You then take a walk in the grid plane. Every time you cross a grid line you draw in a second plane — the "tiling plane" — a line segment of fixed length in the direction of the normal to that grid line. When you have visited all the cells in the grid plane, you will have tiled the tiling plane with rhombi. A density consisting of the sum of delta-functions at the vertices of the tiling can be shown to have the symmorphic space groups $pnmm$, $pnm1$, or pnm, depending on the value of n and the relative phases of the families of lines making up the grid.

When $n = 2^j$, a simple modification of this procedure gives densities with the non-symmorphic space groups $p2^jgm$ [15]. Label alternate lines in each of the 2^{j-1} families of parallel lines as "even" or "odd". Now when you cross a line in the grid plane, orient the line in tiling space a little to the right or left of the normal, depending on whether the grid line is even or odd. A density consisting of a sum of delta-functions at the vertices of the resulting tiling will have the desired non-symmorphic space group.

A somewhat more intricate modification is required to produce real space densities with the symmorphic space groups $pn1m$ or pn [16]. A second set of families of parallel lines is needed, whose wave-vectors are in the lattice generated by the wave-vectors of the first set of families. You can produce the desired tilings by taking the simple walk (drawing the tile edges parallel to the normals to the grid lines) on the more complicated grid plane containing all the lines in both sets of families. A judicious choice of the second set of families and their relative phases gives the desired space groups.

Acknowledgment. My own contributions to the work described here were supported by the National Science Foundation, Grant DMR86-13368 and through the Cornell Materials Science Center, Grant DMR88-18558.

References

1. A. Bienenstock and P. P. Ewald, Acta Cryst **15**, 1253 (1962).
2. D. S. Rokhsar, N. D. Mermin, and D. C. Wright, Phys. Rev. B**35**, 5487 (1987).
3. N. D. Mermin, D. S. Rokhsar, and D. C. Wright, Phys. Rev. Lett. **58**, 2099 (1987).
4. D. S. Rokhsar, D. C. Wright, and N. D. Mermin, Acta Cryst. A**44**, 197 (1988).
5. D. S. Rokhsar, D. C. Wright, and N. D. Mermin, Phys. Rev. B**37**, 8145 (1988).
6. L. Levitov and J. Rhyner, J. Phys. (Paris) **49**, 1835 (1988).
7. F. Gähler, dissertation, ETH No. 8414, Zürich, 1988.

8. T. Janssen, J. Phys. (Paris) **47** C3-85 (1986); Acta Cryst. A**42**, 261 (1986); Phys. Rep. **168**, 55 (1988).
9. T. Janssen and A. Janner, Adv. Phys. **36**, 519 (1987).
10. P. Kramer, J. Phys. (Paris) **47**, C3-75 (1986).
11. S. Alexander, J. Phys. (Paris) **47**, C3-143 (1986).
12. P. Bak, Phys. Rev. Lett. **54**, 1517 (1985); Phys. Rev. B **32**, 5764 (1985); J. Phys. (Paris) **47**, C3-135 (1986).
13. N. W. Ashcroft and N. D. Mermin, *Solid State Physics*, Holt, Rinehart and Winston, New York, 1976, p. 70.
14. N. D. Mermin, D. A. Rabson, D. S. Rokhsar, and D. C. Wright, submitted to Physical Review B.
15. D. A. Rabson, T.-L. Ho, and N. D. Mermin, Acta Cryst. A**44**, 678 (1988).
16. D. A. Rabson, T.-L. Ho, and N. D. Mermin, Acta Cryst. A**45**, 538 (1989).

Part III

Atomic Structures

Progress on the Atomic Structure of Quasicrystals

C.L. Henley

Department of Physics, Cornell University, Ithaca, NY 14853, USA

Abstract. Theoretical and experimental developments of the past three years towards understanding the atomic arrangements in quasicrystal structures are reviewed. Topics mentioned include: the face-centered-icosahedral Al-Cu-TM alloys, models with large three-shell icosahedral clusters, tiling-like quasiperiodic models defined by the cut method, the discovery of large periodic approximant crystals, six-dimensional refinements of diffraction results and their difficulties, improvements in cluster-packing geometries, and the relation of atomic to electronic structure.

1. Introduction

Within about a year of the first quasicrystal publications [1,2] four main chemical classes were known: (1) the Al-TM class, where TM is a transition metal and Al is substituted by other elements, in particular $Al_{80}Mn_{20}$, $Al_{73}Si_{6}Mn_{21}$, and most recently $Al_{65}Cu_{20}Fe_{15}$, [3] which is not only thermodynamically stable, but shows Bragg diffraction indicating long-range order; (2) the AlZnMg class, where the unfilled orbitals are all s or p, in particular $Al_{60}Cu_{10}Li_{30}$ [4] and $Ga_{23}Zn_{40}Mg_{37}$, [5] which are thermodynamically stable but which lose translational correlations on the scale of 1000Å; (3) The PdSiU class, which has not admitted any variation in composition from $Pd_{58}Si_{21}U_{21}$ [6]; (4) the Ti-TM class, [7] composed of transition metals, best represented by $Ti_{63}Mn_{37}$ [8]. Each of these compounds, excepting $i(AlCuFe)$, has space group $P\bar{5}\bar{3}2/m$, the same space group as the 3d Penrose tiling (3DPT) of "Ammann" rhombohedra projected from the 6 dimensional cubic lattice.

In addition, decagonal quasicrystals are reported in the Al-TM class. They have similar quasiperiodic arrangements in 10-fold planes, but have periodicity along the 10-fold axis 12Å in $d(AlMn)$ 16Å in $d(AlFe)$, and 8Å in $d(AlCuCo)$ [9]. Recently, several approximant crystals have been identified which are rational approximants to the decagonal phase: in particular, the long-known $Al_{13}Fe_{4}$ and $T(AlZnMn)$ crystal structures [10] are approximants [11,12] which have revealed the cluster motif and linkages for decagonal phases [13]. Very recently the first equilibrium decagonal quasicrystal was discovered in $d(AlCuCo)$ [14].

2. Inferring structures from approximant crystals

Initially, great progress was made by recognizing that two complicated structures solved long ago, $\alpha(AlMnSi)$ [15,16] and $T(AlZnMg)$ [17,18]

Springer Series in Solid-State Sciences, Vol. 93 Quasicrystals
Editors: T. Fujiwara · T. Ogawa © Springer-Verlag Berlin, Heidelberg 1990

were in fact approximant structures for the Al-TM and the Al-Zn-Mg classes. (Here "approximant" will mean the unit cell is indistinguishable from a fragment commonly occurring in the icosahedral phase.) Each structure is a bcc packing of clusters which have two concentric atomic shells with full icosahedral symmetry around their center. In addition, in each type the clusters are linked together in two different ways, along twofold and threefold symmetry axes. The linkages orient all the clusters the same way, and just the right lengths to allow triangles (and hence a well-connected network) of clusters. In these approximants the network is a bcc lattice; It is easy to build, instead, a network with average icosahedral symmetry [19], since the set of allowed linkage orientations has icosahedral symmetry. Experimental evidence mostly supports the idea that these motifs form the backbone of the structures [20].

The atomic arrangements in the clusters are *different* in the Al-TM class (54 atoms) and Al-Zn-Mg class (44 atoms), so that these fall into different *structural* classes. Furthermore, up to the second shell, there are no other well-packed icosahedral clusters, suggesting quasicrystals of the Ti-TM and PdSiU chemical classes also fall into one of the above two structure classes. Ref. [17] suggested a convenient criterion for guessing the class: let d be the typical atomic diameter and let a_R be the quasilattice constant [21], then

$$a_R/d \approx 1.65 - 1.75 \text{ (Al-TM)}; \quad a_R/d \approx 2.0 \text{ (AlZnMg)} . \qquad (1)$$

This suggests both Ti-TM and PdSiU have the Al-TM type structure.

In a cluster model, the structural description separates into two problems at different scales: [2] (1) the geometry, i.e. the network of cluster centers and linkages (2) the decoration, i.e. rules for placing atoms after one is given a geometry. If the decoration consisted only of placing identical clusters on the centers, then this separation would literally be a factorization, of the Fourier transform into a simple product of the structure factor of the centers ("geometry") times the form factor of the cluster ("decoration"). In reality, a metal structure cannot have large voids; additional, so-called "glue" atoms (about 30% of the total) must occupy sites between clusters. Thus, identification of the basic clusters and linkage rules is only the starting point of a complete structure model.

Third shells with icosahedral symmetry were later noticed around both the $\alpha(AlMnSi)$ cluster [22] (on the body center) and the $R(AlCuLi)$ cluster [23]. The third shells account for most of the "glue" atoms, but inconsistencies may be introduced when they overlap.

3. The Al-TM structure class

We might attempt to describe the Al-TM quasicrystal structure as a decoration of the Ammann rhombohedra of the 3D Penrose tiling, inspired by the $\alpha(AlMnSi)$ structure [16]. This arrangement puts TM atoms on most vertices (and nowhere else). It is similar to a packing of spheres with $\sim$ 1.6 the diameter of an atom; quite reasonable, since Mn-Mn near neighbors are disfavored. This and other quasiperiodic sphere packings are conveniently described by the "cut method", a generalization of the "projection method" [1,2]. A 3D icosahedrally symmetric shape called the "acceptance domain" is defined in the "perpendicular space"; "atomic surfaces" with this cross section oriented in the perpendicular direction are attached to 6D hypercubic lattice vertices, so that a slice intersecting them picks up a delta function representing a site.

The set of all atoms (Al or TM) must also approximate a sphere packing, since atoms have hard cores; this is described by a larger acceptance domain containing the TM domain. Furthermore, to be faithful to local order in $\alpha(AlMnSi)$, the structure must incorporate many of the $Al_{42}TM_{12}$ clusters with empty center sites, which are called "Mackay icosahedra" (MI). This leads us to identify *cluster centers* as a subset of tiling vertices, described by a smaller acceptance domain. Thus we obtain a set of three concentric acceptance volumes, with successively greater radii in perpendicular space (corresponding to greater densities in real space) for the cluster centers, the Mn sites, and the Al sites, respectively, each approximately a sphere packing.

Around each cluster center, this decoration indeed naturally produces the correct second shell (12 Mn + 30 Al) of the MI. However, in the first shell it typically puts only $\sim$ 7 Al atoms on various threefold axes. These must be removed (by cutting away part of the Al acceptance volume) and replaced by 12 Al sites on the fivefold axes [25].

Duneau and Oguey (DO) [26] obtained a significant improvement in the arrangement of the glue atoms. They removed part of of the glue-atom portion of the atomic surfaces attached to integer 6D coordinates, and in its place added an additional "glue" atomic surface attached to the 6D body centers. This is the best quasiperiodic Al-TM model to date; the only unphysical feature is that real atoms would have significant (about 0.25Å) displacements from the ideal tiling sites [16,17].

A separate development was the recognition of a third icosahedrally symmetric shell in the cluster by Fowler et al. [22]; this object was called "double Mackay icosahedron" (DMI) by Yang, [27,28] who independently proposed it. If two DMI clusters are related by a threefold bond, the third shell decorations are inconsistent; thus only the even [29] clusters should become DMI's. Because of this modulation of even and odd clusters, the $\alpha(AlMnSi)$ space group is simple cubic even though the cluster center geometry is a bcc lattice. The same sublattice alternation, extended to Al-TM quasicrystals, must give them the face-centered icosahedral space group [30] $F\bar{5}\bar{3}2/m$, with weak diffraction spots at positions forbidden in $P\bar{5}\bar{3}2/m$. The most highly ordered quasicrystal, $i(AlCuFe)$, [31] shows exactly this diffraction pattern. Indeed, subsequently it was confirmed that $i(AlMn)$ upon annealing develops diffuse scattering at the same spots [32], so it is plausible that this is the minimum energy icosahedral structure for *all* Al-TM quasicrystals.

A dictionary can now be produced roughly equating the Elser-Henley names [16] of atom site classes (Greek letters), to the Duneau-Oguey names [26] ("letters") and the Cahn *et al.*[33] names (Roman numbers) of acceptance domains, listed in order outward from the center:
(i) MI center with 12 Al(α) = "sA" = (no name) [34];
(ii) Mn sites (most tile vertices) = "TR" [35] = Mn(I);
(iii) Al(β) sites (in second MI shell) = Al shell "A" (inner part)=Al(I);
(iv) Al(γ) glue = Al shell "A" (outer part) = Al(II);

(v) Al(δ) glue = shell "sD" = Al(IV), attached to 6D body centers.

The third DMI shell consists of Al(β) atoms in squares around 2-fold axes and Al(δ) atoms on 5-fold axes. Within the DO model, the latter ("sD") sites are at a radius τa_R from the MI center, pretty close to the radius of Al(δ) atoms in $\alpha(AlMnSi)$. The even/odd modulation in $\alpha(AlMnSi)$ can also be considered an Al/vacancy alternation on Al(δ) sites [16]. These sites have mostly icosahedral coordination shells, making them candidates for Mn sites in $i(TiMn)$ [36] or Cu sites in $i(AlCuFe)$. Diffraction on $i(AlCuFe)$ confirms that the modulation of the 6D density (see Sec. 5) does appear near 6D body centers, i.e. on δ sites, but it may be a vacancy/Al or an Al/Cu alternation. [37].

Approximants – Originally, several AlMn phases – large cell crystals λ and μ, and the decagonal phase – were found to have pseudo-icosahedral electron diffraction patterns, suggesting that these phases were also packings of the MI units. [38] However, this was later found to be incorrect for $\mu(AlMn)$ [39]. More recently, truly promising large-cell approximants have been discovered. In the $Al - TM$ family, $i(AlCuFe)$ was found to transform upon cooling to a rhombohedral crystal [40]. Approximants are also observed in $i(TiMn)$ (Ti-TM class) [41].

Other compositions – No structure refinement has yet been undertaken in the Ti-TM class. The Ti_2Ni cubic phase is not a true approximant (no large icosahedral clusters), but it has local order like the Al-TM class, as noticed by Yang [27]. It contains, along the (110) direction, repeating chains of two atoms with icosahedral shells and one with a pentagonal prism. Such a chain is found along the fivefold axis of the three-shell DMI from the central icosahedral vacancy through the pentagonal prism Al(α) atom, and then the icosahedral Mn and Al(δ). So perhaps the $i(TiMn)$ structure is like $i(AlCuFe)$ with Al $\rightarrow$ Ti and (Cu,Fe) $\rightarrow$ Mn.

Because no closely related crystals are known, the structure of the PdSiU class is the least certain of the four chemical classes. The radial distribution function does show that U atoms sit on tile vertices, [42] strongly suggesting an Al-TM type structure with TM $\rightarrow$ U.

4. The AlZnMg structure class

It is now known in the crystalline approximant $R(AlCuLi)$ that (i) the center atom is empty (as in the MI) [43] (ii) the occupation by Li has icosahedral symmetry in the third shell [23]. This suggests that clusters are the physically meaningful unit for packing (Not an obvious point: in contrast to the Al-TM case it *is* feasible here to decorate Penrose tiles directly [17], making a somewhat different model containing more fragments than complete icosahedral clusters.)

The most important recent development is the proposal, independently by Audier *et al.*[44,45] and Ohashi [46], of structure models mixing large and small clusters. Ohashi [46] has pointed out that the solved $Al_5Cu_6Mg_2$ structure [47] is a simple cubic packing of small clusters with short linkages; Audier speculates the same for $Z(AlCuLiMg)$ [45].

With short linkages, it becomes easy to build a deterministic quasiperiodic model by decorating in the same way each large rhombohedron (of edge $\tau^3 a_R$) of the inflated 3D Penrose tiling [44]. This quasiperiodic model is the best AlZnMg class model so far: the atoms appear to be well-packed everywhere with proper chemical order, and – impressively – the *differences* in stoichiometry between the crystal and quasicrystal are correctly predicted. However, much remains to be tested: there is no physical criterion why this particular arrangement is simpler or better than others, since many other locally similar packings are *not* organized into identically decorated large rhombohedra.

New related approximant-like phases have been discovered in the AlZnMg class. A hexagonal Z phase, and two tetragonal phases were found in the AlCuLiMg system [45]. Furthermore, Ohashi [46] has found five new approximant phases in the GaZnMg system, the largest being the "3/2" cubic structure [48] with lattice constant 36.9Å which is inflated by τ^2 from the simple bcc packing of clusters.

5. Six-dimensional analysis

A truly quasiperiodic, icosahedrally symmetric structure can always be represented as a cut through a six-dimensional density. The natural adaptation of conventional crystallographic structure determination is to: index the diffraction peaks as Bragg peaks (using 6D indices), measure their intensities, solve the problem of phases, and Fourier transform into real 6D space [49,50]. A difficulty is that, in principle, an infinite number of parameters is necessary, but nature is kind in that (i) the icosahedral space groups are centrosymmetric $P532/m$ (or $F532/m$) so the phase factors to be determined are ± 1 (ii) much of the atomic structure is similar to a decorated tiling; consequently the Fourier amplitudes, in a first approximation, are smooth functions of the complementary reciprocal lattice vector $\vec{q}^{\perp}$. (This is striking when one plots the intensities as a function of the complementary wavevector $\vec{q}^{\perp}$ [51,52]). Consequently one can guess the phase factors; such a program has been carried out by two different groups for $i(AlMnSi)$ using neutron and X-ray powder diffraction [33,49-52]. The guessed phases are consistent with the *relative* phases between the Mn and Al/Si structure factors for each wavevector, known in $i(AlMnSi)$ from neutron scattering with isomorphous substitution of Mn $\rightarrow$ (Fe,Cr) [49,51].

Both groups' descriptions of the 6D density includes one atomic surface centered on integer vertices in 6D vertices, and just one other atomic surface centered on the body centers. This last is an oversimplification, as is apparent when the model is compared with the DO quasiperiodic model. The central portion of the experimental body-center surface is the same as the DO body-center surface, consisting of Al(δ) atoms. But the outer annulus of the body-center surface corresponds in the DO model to 12 Al(α) surfaces, each one a displaced copy of the MI-center acceptance volume, and these are not in the same 3-plane as the the body center. In real space, using the body center puts the Al(α) atoms on the MI cluster's 5-fold axis between the empty center and Mn at a radius $\tau^{-1} a_R$, which is much too close (1.74Å) to the Mn which is at a_R.

A quasicrystal may be a random packing of clusters, which nevertheless may be long-range ordered due to the rigid local constraints on the cluster linkages (random tiling scenario, see Sec. 6). In such a case, the 6D approach is seriously incomplete. The Fourier transform of the Bragg intensities gives only an averaged structure, in which *most* sites are expected to be partially occupied. (Exactly this is found in the 3D densities reconstructed from experiments [51]; a somewhat different explanation for the spurious features in real space is in Ref. [53]).

This is inadequate as a structure model, because when we have two nearby partial atoms, we do not know whether both are present at the same time, or whether they alternate with each other. A true structure model must describe an ensemble of typical realizations, each of which is a reasonable atomic structure; such information is usually necessary in order to calculate physical properties (magnetism, electronic structure).

6. Cluster packing geometries

Implicit in the cluster idea is that grouping into clusters with linkages at special orientations is energetically favorable. Thus we expect the clusters to be packed as tightly as possible, consistent with the constraints of icosahedral symmetry and the allowed kinds of linkages. This leads us to try constructing a well-packed quasiperiodic network of cluster centers, parameterized (as in Sec. 2) as a cut through a 6D cubic lattice, where each vertex has attached a copy of the 3D acceptance volume, which is adjusted to optimize the packing [24, 52]. The best such networks known so far [24,26], with spheres placed on the centers, attain the rather miserable packing fraction $f \approx 0.55$. A random growth model with the same linking constraints [19] attains $f \approx 0.62$, about as good as random-close-packing; surely a quasiperiodic packing can do equally well.

A significant recent development in cluster models is the proposal of a network built of both large (three-shell) and small (two-shell) fundamental clusters. Besides the same twofold and threefold axis linkages between large-large (overlapping) or large-small clusters, there is a new twofold small-small linkage shorter by a factor τ^{-1} than the previously known 2-fold linkage [28,44-46].

This makes it easy to construct a quasiperiodic model, with highly symmetrical, self-similar arrangements of the clusters. However, this is unsatisfying physically: there are many other differently organized packings with the same statistics of clusters and linkages, but so far there is no physical criterion to select among them.

A great deal of controversy and study has focused on what sort of disorder causes the peak broadening seen in all quasicrystals except $i(AlCuFe)$, and whether the ground state is a true quasiperiodic crystal. This concerns the geometry and is outside the scope of this paper. However, it is relevant to mention that the known atomic decorations are all too symmetric to enforce the "matching rules" that are the only plausible way to make a quasiperiodic ground state of an alloy. Nonequilibrium random growth models (describing rapidly quenched materials) are reviewed elsewhere [19,54]. The related "random tiling models", are a class of *equilibrium* geometry models [55]. The structure has a quasicrystal phase - with Bragg diffraction – at higher temperatures (stabilized by configurational entropy) but should transform to a crystal at low temperatures [56]. This seems to be consistent with some of the data on *i(AlCuFe)* [56,57].

Canonical cell model – So far, atomic structure models based on random packings of clusters have always used ad-hoc, nondeterministic ways to place "glue" atoms between clusters. Since these spaces are small, we expect the glue atom positions should rather be strictly determined by the surrounding cluster geometry. If neighboring clusters and their edges form polyhedra, and we constrain these to a limited set of the best-packed possibilities, then it is possible to formulate a model where (1) each unit (cluster or polyhedron) has a unique decorations and (2) the atoms decorating distinct units are always disjoint (no sharing). In that case, the Fourier transform *does* become a sum (over different types and orientations of units) of products of form factors and structure factors. The polyhedra are called "canonical cells" and the structure is in fact a tiling of these cells. Both random and quasiperiodic packings of the cells probably exist, with packing fractions $f \approx 0.62$, but to date only periodic approximant models have been constructed [48].

7. Electronic versus atomic structure

Electronic structure calculations have begun to address the structure questions. It was shown within the "embedded atom" approximation that Mn atoms prefer a local enviroment with a higher electron density, as in an icosahedral coordination shell [58].

Indeed, the "double" MI cluster is as a combination of 12 vertex-sharing $MnAl_{12}$ icosahedra [27,28,58]. An alternative approach, applied to Ti-TM, finds angular four-body effective interatomic potentials that favor tetrahedra around TM and octahedra of Ti (as also formed by Al in Al-TM structures) [36]. In fact, many Al-TM phases contain the structural motif of a large tetrahedron of TM atoms, each icosahedrally coordinated, with Al atoms on the edges (forming an octahedron). These tetrahedra pack just like the large tetrahedral "Friauf polyhedron" found in Frank-Kasper phases related to the AlZnMg class [59]; the respective Al-TM and AlZnMg icosahedral clusters can each be viewed as a packing of 20 large tetrahedra.

It is still uncertain whether there is any electronic basis for treating the clusters as rigid, more tightly bound entities; there is a small hint in a recent large-scale bandstructure calculation for "$\alpha(AlMn)$" where states were found nearly localized on the MI [60].

Most quasicrystals are ternaries, yet the structural models are mostly binary models: (Al,Si)Mn, (Al,Cu)Fe, (Pd,Si)U, or (Al,Cu)Li. Are the extra species important in order for optimal occupation of certain site types? The Hume-Rothery mechanism was suggested,which would imply a "no" answer, i.e. they are just "dopants" to change the conduction electron density [60-62], and might well be randomly mixed. This might explain the differing stoichiometries of $i(AlCuLi)$ and $i(AlZnMg)$ [63].

Acknowledgment. I thank Marc Audier, Veit Elser, Marko Jarić, Ken Kelton, Wataru Ohashi, Rob Phillips, Arthur P. Smith and Ed Stern for discussions. This was supported by the U.S. DOE grant DE-FG02-89ER45405.

References

1. See *The Physics of Quasicrystals*, ed. P. J. Steinhardt and S. Ostlund (World Scientific, 1987).

2. See C. L. Henley, Comments Cond. Matt. 13, 59 (1987).

3. A. P. Tsai, A. Inoue, and T. Masumoto, J. Mater. Sci. Lett. 7, 322 (1988).

4. B. Dubost, J. M. Lang, M. Tanaka, P. Sainfort, and M. Audier, Nature 324, 48 (1986).

5. W. Ohashi and F. Spaepen, Nature 330, 555 (1987).

6. S. J. Poon, A. J. Drehman, and K. R. Lawless, Phys. Rev. Lett. 55, 2324 (1985).

7. Z. Zhang, H. Q. Ye, and K. H. Kuo, Phil. Mag. A52, L49 (1985).

8. K. F. Kelton, P. C. Gibbons, and P. N. Sabes, Phys. Rev. B38, 7810 (1988).

9. L. X. He, Y. K. Wu, and K. H. Kuo, J. Mater. Sci. Lett. 7, 1284 (1988), and references therein.

10. P. J. Black, Acta Cryst. 8, 175 (1955); K. Robinson, Acta Cryst. 7, 494 (1954); A. Damjanović, Acta Cryst. 14, 982 (1961).

11. C. L. Henley, J. Non-Cryst. Solids 75, 91 (1985).

12. G. van Tendeloo, J. van Landuyt, S. Amelinckx, and S. Ranganathan, J. Microsc. 149, 1 (1988).

13. C. L. Henley, unpublished.

14. A. R. Kortan, F. A. Thiel, H. S. Chen, A. P. Tsai, A. Inoue, and T. Masumoto, Phys. Rev. B40,9397 (1989).

15. P. Guyot and M. Audier, Phil. Mag. 52, L15 (1985).

16. V. Elser and C. L. Henley, Phys. Rev. Lett. 55, 2883 (1985).

17. C. L. Henley and V. Elser, Phil. Mag. 53, L59 (1986).

18. $R(AlCuLi)$ is almost isostructural to $T(AlZnMg)$.

19. V. Elser in *Aperiodic Crystals III*, ed. M. V. Jarić (Academic Press, New York, 1989).

20. Y. Ma, E. A. Stern, and F. W. Gayle, Phys. Rev. Lett. 58, 1956 (1987); Y. Ma and E. A. Stern, Phys. Rev. B38, 3754 (1988).

21. The quasilattice constant a_R is conveniently defined as the rhombohedron edge length (roughly 5$\AA$ in all cases) of the tiling with the same reciprocal lattice (see Refs. 1-2). Note also we must take d to be the average neighbor distance *in structures of similar chemistry.*

22. H. A. Fowler, B. Mozer, and J. Sims, Phys. Rev. B37, 3906 (1988).

23. M. Audier, P. Sainfort, and B. Dubost, Phil. Mag. B54, L105 (1986).

24. C. L. Henley, Phys. Rev. B34, 797 (1986).

25. A. Yamamoto and K. Hiraga, Phys. Rev. B37, 6207 (1988). The Al_{12} icosahedron is best considered as a single object decorating the MI center site.

26. M. Duneau and C. Oguey, J. Phys. (France) 50, 135 (1989).

27. Q. B. Yang, Phil. Mag. Lett. 57, 171 (1988).

28. Q. B. Yang, Phil. Mag. B58, 47 (1988).

29. One can divide the cluster network into even and odd, like a bipartite lattice, so neighbors related by a threefold linkage have opposite parities.

30. C. L. Henley, Phil. Mag. Lett. 58, 87 (1988).

31. S. Ebalard and F. Spaepen, J. Mater. Res. 4, 39 (1989).

32. N. K. Mukhopadhyay, S. Ranganathan, and K. Chattopadhyay, Phil. Mag. Lett. 60, 207 (1989).

33. J. W. Cahn, D. Gratias, and B. Mozer, J. Phys. (France) 49, 1225 (1988).

34. In Ref. 33, the Al(α) are assigned as part of a body center decoration = Al(III), but this is an oversimplification (see Sect. 6).

35. More precisely, Mn are in "TR" minus the MI center and Al(β) domains.

36. R. B. Phillips, Ph. D. thesis (Washington University, St. Louis, 1989).

37. D. Gratias, B. Mozer, Y. Calvayrac, J. Devaud-Rzepski, and A. Quivy, unpublished.

38. L. Bendersky, J. Microsc. 146, 303 (1987).

39. C. B. Shoemaker, D. A. Keszler, and D. P. Shoemaker, Acta. Cryst. B45, 13 (1989).

40. M. Audier and P. Guyot, in *Quasicrystals and Incommensurate Structures in Condensed Matter*, ed. M. J. Yacaman, D. Romeu, V. Castaño, and A. Gómez (to appear, World Scientific, 1990).

41. L. Levine, K. Kelton, P. Gibbons, and J. Holtzer, unpublished.

42. D. D. Kofalt, I. A. Morrison, T. Egami, S. Preische, S. J. Poon, and P. J. Steinhardt, Phys. Rev. B35, 4489 (1987).

43. S. Samson (unpublished); C. A. Guryan, P. W. Stephens, A. I. Goldman, and F. W. Gayle, Phys. Rev. B37, 8495 (1988); M. Audier et al, Phil. Mag. B153, 136 (1988).

44. M. Audier and P. Guyot, unpublished

45. M. Audier, Ch. Janot, M. de Boissieu, and B. Dubost, Phil. Mag. B60, 437 (1989);

46. W. Ohashi, Ph. D. thesis (Harvard University, 1989).

47. S. Samson, Acta. Chem. Scand. 3, 809 (1949); see Structure Reports 12, 8 (1949).

48. C. L. Henley, unpublished (canonical cells).

49. See review in C. Janot, M. De Boissieu, J. M. Dubois, and J. Pannetier, J. Phys. Condens. Matt. 1, 1029 (1989).

50. An excellent discussion is in D. Gratias, J. W. Cahn, M. Bessière, Y. Calvayrac, S. LeFebvre, A. Quivy, and B. Mozer, in *Fractals, Quasicrystals, Chaos, Knots, and Algebraic Quantum Mechanics*, ed. A. Amann, L. Cederbaum, and W. Gans (Kluwer Academic, 1988).

51. C. Janot, J. Pannetier, J. M. Dubois, and M. De Boissieu, Phys. Rev. Lett. 62, 450 (1989).

52. J. W. Cahn and D. Gratias, J. Phys (France) Colloq. 47, C3-415 (1986).

53. Y. Ma, E. A. Stern, X.-O. Li, and C. Janot, Phys. Rev. B11, 8053 (1989).

54. P. W. Stephens in *Aperiodic Crystals III*, ed. M. V. Jarić (Academic Press, New York, 1989).

55. C. L. Henley, J. Phys. A. 21, 1649 (1988); M. Widom, D. P. Deng, and C. L. Henley, Phys. Rev. Lett. 63, 310 (1989); K. H. Strandburg, L. H. Tang, and M. V. Jarić, Phys. Rev. Lett 63, 314 (1989).

56. C. L. Henley, in *Quasicrystals and Incommensurate Structures in Condensed Matter*, ed. M. J. Yacaman, D. Romeu, V. Castaño, and A. Gómez (to appear, World Scientific, 1990); M. Widom, in *Quasicrystals*, ed. M. V. Jarić and S. Lundquist (World Scientific, in Press, 1989).

57. P. A. Bancel, preprint.

58. A. C. Redfield and A. Zangwill, Phys. Rev. Lett. 58, 2322 (1987).

59. S. Samson, in *Structural Chemistry and Molecular Biology*, ed. A. Rich and N. Davidson (Freeman, San Francisco, 1968).

60. T. Fujiwara, Phys. Rev. B40, 942 (1989).

61. J. Friedel, Helv. Phys. Acta 61, 538 (1988).

62. The idea is that structures are especially stable for a conduction electron density such that the (nearly-free-electron) Fermi sphere diameter is close to a reciprocal lattice vector with a large structure factor, since the splitting of states near the Fermi level by the lattice potential raises the energy of empty states and lowers that of filled ones.

63. V. G. Vaks, V. V. Kamyshenko, and G. D. Samolyuk, Phys. Lett. A132, 131 (1988).

From Crystal Approximants to Quasicrystals

M.V. Jarić [1,2] *and Shi-Yue Qiu* [2]

[1] Department of Physics, University of California, Santa Cruz, CA 95064, USA

[2] Center for Theoretical Physics, Texas A&M University, College Station, TX 77843, USA

Abstract. It is proposed that the phase problem for quasiperiodic quasicrystal structures can be solved by identifying a rational approximant of the quasicrystal with a crystal of a known structure. Such identification leads to the constraints on the unknown phases of the quasicrystal density waves. In favorable cases, these constraints can uniquely fix the phases of the density waves whose intensities are measurable. The method is illustrated on quasicrystalline i(Al-Cu-Li). The 6D quasicrystalline electron density so determined suggests a hollow vertex and a solid mid-edge atomic surfaces.

1. Introduction

Icosahedral quasicrystals are experimentally defined by sharp diffraction peaks indexed by a six-component icosahedral basis. Neither the icosahedral symmetry, which determines the orientational order, nor the sharpness of the peaks, which determines the extent of the positional order, are necessarily perfect in real quasicrystals. Both kinds of order are idealized (long-ranged) only for a quasiperiodic crystal. A quasiperiodic crystal can be regarded as a cut through a higher-dimensional crystal whose "atoms" are surfaces of codimension three [1]. Provided the atomic surfaces are simple, quasiperiodicity can offer a meaningful, even if approximate, description of the structure of quasicrystals. It is certainly a good assumption for i(Al-Fe-Cu), with positional correlation length $\xi \geq 10^4$Å, and probably so for i(Al-Li-Cu) ($\xi \sim 10^3$Å). It is less certain for i(Al-Mn-Si) ($\xi \sim 10^2$Å) where both kinds of order are less perfect. Disorder, either thermal or quenched, has a role of smearing the atomic surfaces into a smooth density.

As it is well known, diffraction experiments provide the amplitudes of the density waves, but not their phases. While general methods for solving this "phase problem" have been developed for crystals, they are unknown for quasiperiodic crystals. It has been recognized [2] that, very much like in the case of modulated (incommensurate) crystals, a quasicrystal is often structurally related to a crystal. This relationship can be exploited, for example, by repacking certain atomic clusters from the crystal [2] into a prescribed quasiperiodic geometric network, such as the Ammann tiling, or by lifting the crystal into a six-dimensional (6D) hypercrystal and extrapolating thus obtained discrete set of point-like atoms into 3D atomic surfaces characterized by a small number of parameters [3]. It has been difficult to achieve a good agreement with experimental

Springer Series in Solid-State Sciences, Vol. 93 **Quasicrystals**
Editors: T. Fujiwara · T. Ogawa

diffraction data using such approaches. Instead, we present here a new method in which the experimentally obtained diffraction intensities are used as an input in determining the phases of the density waves of the quasiperiodic crystal. The phases are obtained by the best fit between a rational approximant (a uniformly phason-strained quasicrystal) and the related crystal. In this way a quasicrystal structure, smeared by quenched or thermal disorder present in the real quasicrystal, can be obtained.

We shall briefly explain the necessary theoretical background and then present the results for icosahedral Al-Li-Cu. More details can be found elsewhere [4].

2. The Method

A uniform phason strain of a quasicrystal can be viewed as a simple shear of the associated hypercrystal. A shear that distorts the hypercrystal into a rational relationship with the physical space corresponds to a phason strain that distorts the quasicrystal into a crystal, the so-called rational approximant. In the reciprocal hyperspace, a uniform phason strain $\epsilon^{\parallel,\perp}$ leaves values of the Fourier components of the density unchanged, but it shifts the reciprocal wavevectors in proportion to their "phason" components $\mathbf{Q}^{\perp}$,

$$\mathbf{Q} \rightarrow \mathbf{Q}_r = \mathbf{Q} - \epsilon^{\parallel,\perp} \cdot \mathbf{Q}^{\perp}. \tag{1}$$

The dense set $\{\mathbf{Q}\}$ of a quasicrystal collapses into a periodic lattice $\{\mathbf{Q}_r\}$ for special values of $\epsilon^{\parallel,\perp}$ indicating a distortion from the quasiperiodic structure to a rational approximant. For example, for i(Al-Li-Cu) a family of rational approximants can be obtained by choosing

$$\epsilon^{\parallel,\perp} = \frac{q\tau - p}{p\tau + q} I, \tag{2}$$

where p and q are integers, and I is the 3×3 unit matrix. The generalized Miller indices [5] map under this distortion as

$$(h/h', k/k', l/l') \rightarrow (h_r, k_r, l_r) = (qh + ph', qk + pk', ql + pl'), \tag{3}$$

while the hypercubic lattice constant $\overline{a}$ maps into the cubic lattice constant

$$a_r = \sqrt{2/(2+\tau)}(q + p\tau)\overline{a}. \tag{4}$$

The best approximants to the quasiperiodic structure are obtained for $p/q = F_{n+1}/F_n$ in the limit of large n, where F_n is the nth Fibonacci number. The R(Al-Li-Cu) crystal is closest to $p/q = 1/1$ bcc rational approximant.

The structure of a rational approximant is not unique even for a fixed $\epsilon^{\parallel,\perp}$. The quasiperiodic structure is changed only trivially by a phase $\mathbf{u}^{\perp} \cdot \mathbf{Q}^{\perp}$ which is proportional to $\mathbf{Q}^{\perp}$. However, structure of the approximant depends on such a phase and, consequently, the phason translation $\mathbf{u}^{\perp}$ of the hypercrystal must be taken as a fitting parameter. The required symmetry of the approximant may impose strong restrictions on $\mathbf{u}^{\perp}$. For example, the space group $Im3$ of the R-phase forces

$$\mathbf{u}^{\perp} = \mathbf{0} \text{ or } \sqrt{2(2+\tau)}\bar{a}(1,1,1), \tag{5}$$

both choices leading to the same structure.

The Debye-Waller factors associated with the "phason" fluctuations of the atomic surfaces in directions perpendicular to the physical space [6], are inherited by the rational approximant but do not exist in a real crystal and must be removed. Since the phason Debye-Waller factors of the quasicrystal are not known, they must be taken as fitting parameters. As a first step, only one such factor, $\exp(-B|\mathbf{Q}^{\perp}|^2)$, can be introduced.

Based on the above considerations, apart from a possible trivial spatial rescaling, the Fourier components of the rational approximant $\rho_r(\mathbf{Q}_r)$ and the quasicrystal $\rho(\mathbf{Q}, \mathbf{Q}^{\perp})$,

$$\rho(\mathbf{Q}, \mathbf{Q}^{\perp}) \equiv \sqrt{I(\mathbf{Q}, \mathbf{Q}^{\perp})} \exp i\theta(\mathbf{Q}, \mathbf{Q}^{\perp}), \tag{6}$$

are related by

$$\rho_r(\mathbf{Q}_r) = \sum_{\mathbf{Q}^{\perp}(\mathbf{Q}_r)} \rho(\mathbf{Q}_r + \epsilon^{\|,\perp} \cdot \mathbf{Q}^{\perp}, \mathbf{Q}^{\perp}) \exp(i\mathbf{u}^{\perp} \cdot \mathbf{Q}^{\perp} + B|\mathbf{Q}^{\perp}|^2), \tag{7}$$

where $I(\mathbf{Q}, \mathbf{Q}^{\perp})$ is the measured diffraction intensity, and for a given $\mathbf{Q}_r$ the sum extends over infinitely many $\mathbf{Q}^{\perp}$. The difference between the rational approximant $\rho_r(\mathbf{r})$ and the crystal $\rho_c(\mathbf{r})$ can be minimized by simultaneously minimizing $|\rho_r(\mathbf{Q}_r) - \rho_c(\mathbf{Q}_r)|^2$ for all $\mathbf{Q}_r$. Therefore, infinitely many phases $\theta(\mathbf{Q}_r + \epsilon^{\|,\perp} \cdot \mathbf{Q}^{\perp}, \mathbf{Q}^{\perp})$ must be determined from a single condition at each $\mathbf{Q}_r$. However, due to a finite experimental resolution, for many $\mathbf{Q}_r$ the sum in Eq. (7) will have only a contribution from the smallest $|\mathbf{Q}^{\perp}|$. The phase factors can be determined directly for such peaks. Even when more than one term contributes to the sum, a unique solution can be often found since the allowed values for the phases are often discrete. Also, more than one related crystal structure may be available, increasing the number of constraints. However, even when there is only one related crystal structure, all the constraints Eq. (7) will not be consistent in general, and one may choose to minimize

$$R_\rho = \frac{\sum_{\mathbf{Q}_r} |\rho_r(\mathbf{Q}_r) - \rho_c(\mathbf{Q}_r)|^2}{\sum_{\mathbf{Q}_r} |\rho_c(\mathbf{Q}_r)|^2}, \tag{8}$$

a quantity analogous to the usual reliability factor.

3. Application to i(Al-Li-Cu)

A good candidate for an application of this new method is icosahedral quasicrystalline i(Al-Li-Cu) whose crystalline counterpart is R(Al-Li-Cu). Their compositions ($Al_{0.570}Cu_{0.108}Li_{0.322}$ vs. $Al_{0.564}Cu_{0.116}Li_{0.32}$) and densities (2.47 vs. 2.46 g cm^{-3}) are very similar [7], the lattice constants of the rational approximant and the real crystal (13.92 vs. 13.90Å) are close, and the structure of R(Al-Li-Cu) is well known [8]. Moreover, the

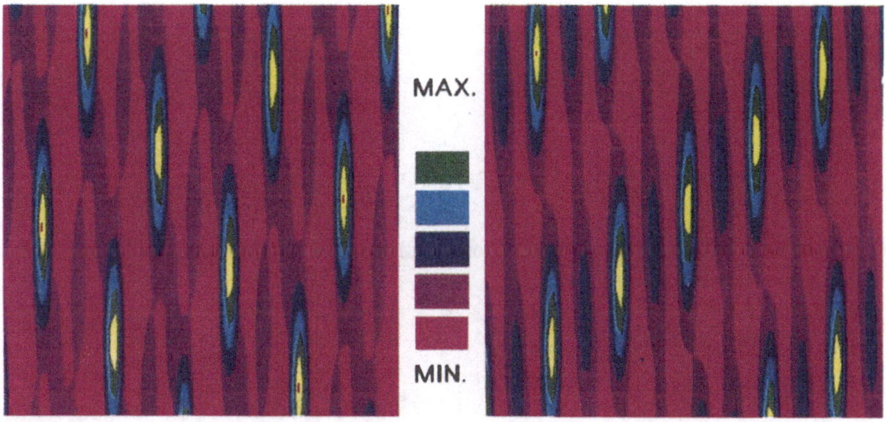

Figure 1. The quasiperiodic Patterson function of quasicrystalline i(Al-Li-Cu) is represented as the six-dimensional periodic Patterson function shown here in 30Å×30Å sections of two-dimensional two-fold and five-fold symmetry planes. The horizontal axis is in the physical space.

6D Patterson function of i(Al-Li-Cu), shown in Fig. 1 is simple, so that a quasiperiodic model of i(Al-Li-Cu) is probably meaningful. Single-grain crystal and quasicrystal X-ray data were collected and a standard refinement of the crystal structure performed at ILL in Grenoble [4]. For R(Al-Li-Cu) crystal, 6099 reflections, or 712 independent reflections, with the maximum (h, k, l) Miller indices (25, 25, 25) were determined. For i(Al-Li-Cu) quasicrystal, about 1000 reflections with intensities larger than three times their standard deviations were measured, corresponding to 45 independent reflections. Using the indexing with the six Miller pseudo-cubic indices $(h/h', k/k', l/l')$ and the 6D hypercubic lattice constant $\overline{a} = 7.15$Å, every diffraction peak could be indexed by

$$\mathbf{Q} = \frac{2\pi}{\overline{a}\sqrt{2(2+\tau)}}(h + h'\tau, k + k'\tau, l + l'\tau), \qquad (9)$$

where $\tau = \frac{1}{2}(1+\sqrt{5})$ is the golden mean. They correspond to $N \leq 108$ and $M \leq 172$ [5], exhibit allowed point symmetries, and belong to a primitive icosahedral Bravais lattice.

For i(Al-Li-Cu) most of the phases could be determined uniquely. The remaining phases and the constant $B = 1.24$ were determined by minimizing R_ρ. The remarkable agreement between the 1/1 rational approximant and the crystal shown in Fig. 2 is obtained under the assumption that the quasicrystal has a symmorphic, centrosymmetric hyperspace group. The resulting phase factors are shown in Table 1. Only Li, which is practically invisible to X-rays, and probably buried in the quasicrystal diffraction background, is missing from the rational approximant. Using the phases which led to the rational approximant in Fig. 2, we can reconstruct the quasicrystal electron density. We show in Fig. 3 projections of 50Å thick slices of the quasicrystal perpendicular to 2-, 3-,

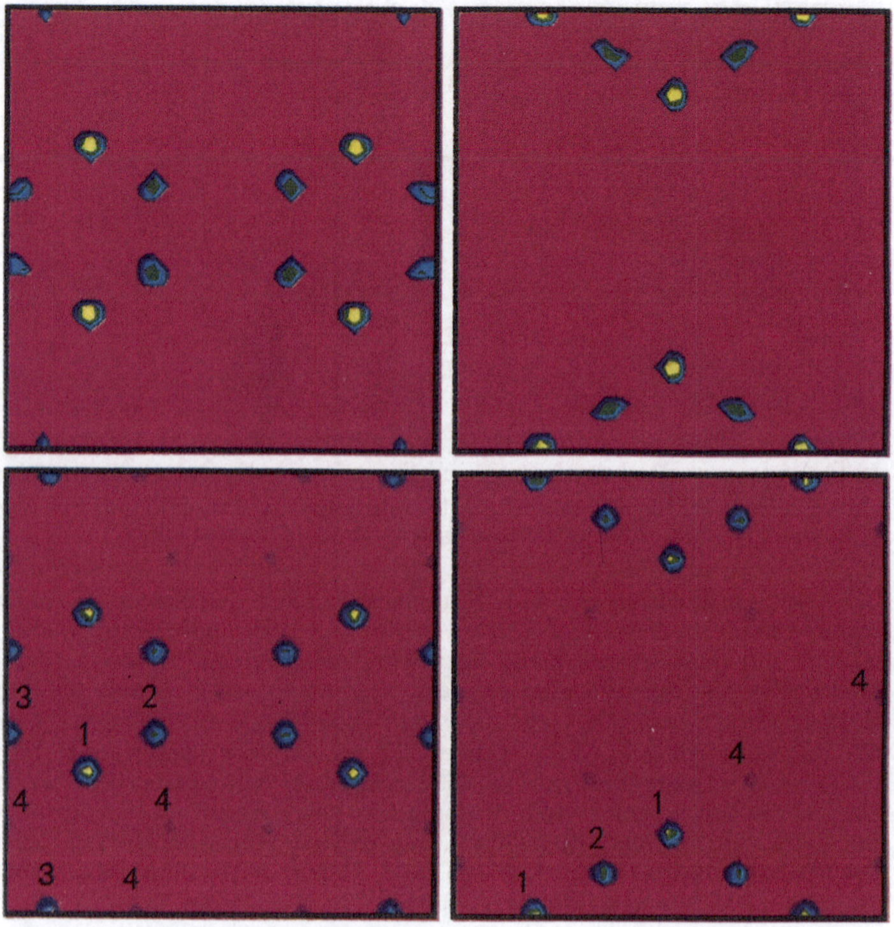

Figure 2. The electron density of the 1/1 rational approximant of i(Al-Li-Cu) shown in row one was obtained for the choice of phases which maximizes its agreement with the electron density of R(Al-Li-Cu) shown in row two. Cuts perpendicular to the four-fold rotation axis through the center of symmetry and at a distance 0.191 of the lattice constant are shown in columns one and two, respectively. The numbers denote: 1=$Cu_{46.2}Al_{53.8}$, 2=$Cu_{11.4}Al_{88.6}$, 3=Al, and 4=Li.

and 5-fold rotation axes. They compare favorably with the actual transmission electron micrographs of i(Al-Li-Cu). Also, the 6D hypercrystal electron density can be constructed. As illustrated in Fig. 4, where high symmetry 2D rational sections through the hypercrystal are shown, the 6D periodic density is mostly concentrated in two types of narrow 3D regions perpendicular to the physical space and centered at the vertices and edges of the hypercubic lattice. Projections of the density within a narrow region of the physical space onto the 3D perpendicular space centered at a mid-edge and a vertex are shown in Fig. 5 (a) and (b), re-

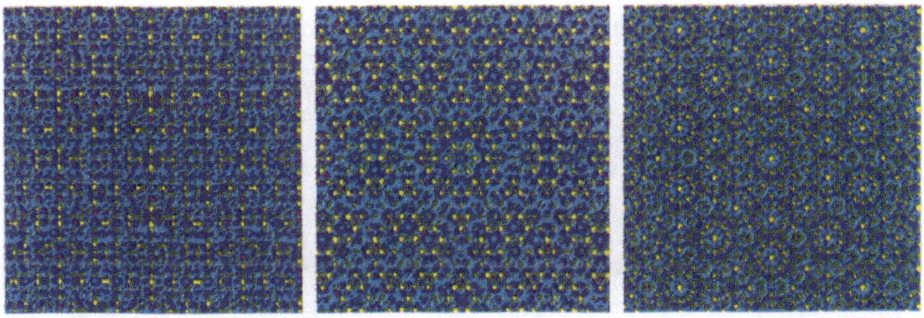

Figure 3. Calculated projections of i(Al-Li-Cu) electron density on the physical planes perpendicular to 2-, 3-, and 5-fold rotation axes. The projection thickness of the shown 60Å×60Å slices is 50Å.

Table 1. Measured diffraction intensities for i(Al-Li-Cu) (arbitrary units) and associated phase factors determined from the fit shown in Fig. 2.

$I(\mathbf{Q})$	$e^{i\theta}$	$\mathbf{Q}$	$I(\mathbf{Q})$	$e^{i\theta}$	$\mathbf{Q}$
1.87	1	0 0 1 0 0 1	3.23	-1	1 2 3 1-2 2
26.10	-1	1 0 1 1 0 0	78.82	1	0 2 3 0-2 3
2.05	-1	1 1 1 0 0-1	1.80	-1	3 2 1 1 3-2
2.66	-1	0 1 1 0-1 1	8.03	-1	3 1 3 3-1 0
1.77	1	1 1 1 1-1 0	14.75	-1	3 1 3 3-1-1
1.20	-1	2 1 0 1 0 0	13.40	1	1 2 4 1-2 2
2.73	-1	1 1 1 1-1 1	2.20	1	3 0 3 3 0-2
1.30	-1	2 0 1 2 0 0	1.18	-1	4 3 2 1 1-1
39.83	1	1 1 2 1-1 1	7.17	1	0 2 4 0-2 3
1.50	1	1 2 2 0 1 0	3.47	1	2 2 4 2-2 1
95.59	1	0 1 2 0-1 2	2.06	1	1 2 4 0-2 3
1.15	1	2 1 1 2 0-1	5.07	-1	1 2 4 1-2 3
1.48	1	2 0 2 2 0 0	99.90	1	2 2 4 2-2 2
60.89	-1	2 0 2 2 0-1	1.00	-1	4 3 1-1 3-1
2.35	-1	2 2 2-1 1 0	1.60	-1	3 2 4 3-1 0
23.15	1	1 1 3 1-1 1	18.55	1	0 4 0-4 2 2
10.31	-1	2 1 2 2-1 0	27.43	-1	3 4 0-2 4 0
1.45	-1	0 1 3 0-1 2	7.68	-1	4 1 3 4 0-2
3.12	-1	2 1 2 2-1-1	2.18	1	1 3 5 1-2 3
1.19	1	1 2 3 1 0 1	1.37	-1	4 1 4 4 0-1
3.26	1	1 1 3 1-1 2	2.70	-1	4 1 4 4-1-1
9.63	1	2 1 3 2-1 0	1.10	1	4 2 4 4-1 0
			1.20	-1	5 1 1 3 3-3

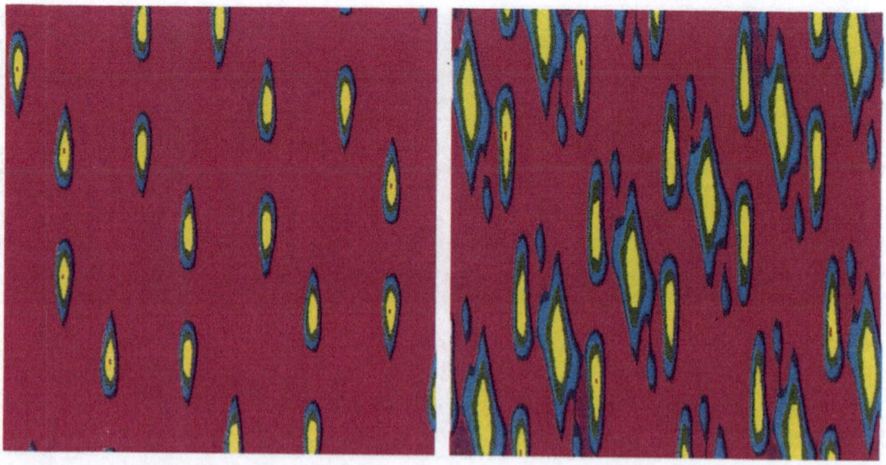

Figure 4. Electron density of i(Al-Li-Cu) quasicrystal lifted into the six-dimensional hypercubic lattice and shown in 30Å×30Å 2-fold and 5-fold rotation planes. The horizontal axis is in the physical space.

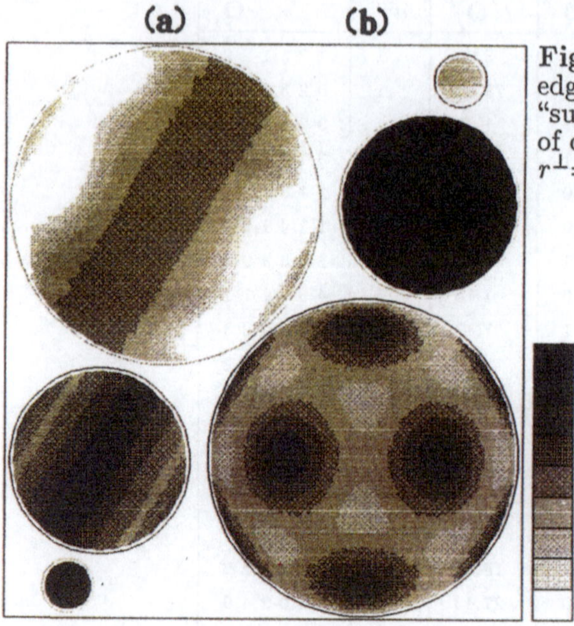

Figure 5. The smeared mid-edge (a) and vertex (b) atomic "surfaces" are shown in terms of densities on spheres of radii $r^{\perp}$=1Å, 4Å, and 7Å.

spectively. The densities are shown on spheres of radii $r^{\perp}$ =1Å, 4Å, and 7Å. It appears that the atomic surfaces do not have a significant physical-space component. Their smearing is caused by a phason disorder which may also involve a chemical Al/Cu disorder. The determination of the precise shape and composition of the atomic surfaces is beyond the scope of the present work.

4. Conclusions

In summary, we presented a new method for solving the phase problem in quasicrystals using a similarity between rational approximants and crystals of known structure. As an illustration we applied this new method to i(Al-Li-Cu). A direct comparison in Fig. 2 of the electron densities of the 1/1 approximant and the crystalline R(Al-Li-Cu) shows a remarkable agreement, except for the missing Li which is probably smeared by the phason disorder. The hypercrystal electron density shown in Fig. 4 reveals smeared atomic surfaces at mid-edges and vertices of the hypercubic lattice. They are shown in Fig. 5 where the phonon smearing is removed by integration. The vertex surface of Y_h symmetry is hollow, while the edge-center surface of D_{5v} symmetry is solid. Part of the phason smearing of the atomic surfaces which remains in Fig. 5 should be removed by introducing separate phason Debye-Waller factors for each surface and by modeling the experimental truncation and disorder present in the quasicrystal. Already collected partial structure factors [9] will be used to specify the Li surfaces and the chemical content of the Al/Cu surfaces. The chemical disordering problem can be avoided in materials like i(Al-Cu-Fe) which are apparently chemically ordered. This method can be used as a starting point for quasicrystal structure refinement and specification of the atomic surfaces.

5. Acknowledgments

We are grateful to S. L. Johnson for computer assistance. This work was supported in part by the NSF grants No. DMR8821802 and DMR8721673, by the Nonlinear Science Institute at the University of California Santa Cruz, and by the Regents of Texas A&M University through the Materials Science and Engineering Program and the Center for Theoretical Physics.

6. References

[1] P. Bak, Phys. Rev. Lett. **56**, 861 (1986); Phys. Rev. Lett. **54**, 1517 (1985); Phys. Rev. B **32**, 5765 (1985).

[2] C. L. Henley, Comm. Cord. Mat. Phys. **13**, 59 (1987); V. Elser, and C. L. Henley, Phys. Rev. Lett. **55**, 2883 (1985); P. Guyot, and M. Audier, Phil. Mag. Lett. B **52**, L15 (1985); C. L. Henley, J. of Non-Cryst. Sol. **75**, 91 (1985); M. Audier and P. Guyot, Proc. ILL/Codest Workshop, ed. C. Janot and J. M. Dubois (World Scientific, Singapore, 1988) p. 181; M. Audier, P. Sainfort, and B. Dubost, Phil. Mag. Lett. B **54**, L105 (1986); J. W. Cahn, and D. Gratias, J. Phys. Colloq. **47**, 415 (1986); P. Guyot, M. Audier, and R. Lequette, J. Phys. (Paris) Colloq. **47**, 389 (1986).

[3] J. W. Cahn, D. Gratias, and B. Mozer, Phys. Rev. B **38**, 1638 (1988); D. Gratias, J. W. Cahn, and B. Mozer, Phys. Rev. B **38**,

1643 (1988); D. Gratias, Proc. ILL/Codest Workshop, ed. C. Janot and J. M. Dubois (World Scientific, Singapore, 1988) p. 83.

[4] S.-Y. Qiu, M. V. Jarić, C. Janot, and M. de Boissieu, Phys. Rev. Lett. (to be published); S.-Y. Qiu, and M. V. Jarić, Proc. Anniversary Adriatico Research Conference on Quasicrystals, ed. M. V. Jarić and S. Lundquist (World Scientific, Singapore, at press); S.-Y. Qiu and M. V. Jarić (unpublished); C. Janot and M. de Boissieu (unpublished).

[5] J. W. Cahn, D. Shechtman and D. Gratias, J. Mater. Res. **1**, 13 (1986).

[6] M. V. Jarić and D. R. Nelson, Phys. Rev. B **37**, 4458 (1988).

[7] M. Audier, C. Janot, M. de Boissieu, and B. Dubost, Phil. Mag. B**60**, 437 (1989).

[8] E. E. Cherkashin, Sov. Phys. Cryst. **8**, 681 (1964); M. Audier, J. Pannetier, M. Leblanc, C. Janot, J. Lang, and B. Dubost, Physica B **153**, 136 (1988); P. Sainfort, B. Dubost, and A. Dubus, Compt. Rend. Acad. Sci. Paris, 689 (1985).

[9] M. de Boissieu, C. Janot, J. M. Dubois, M. Audier and B. Dubost, J. de Phys. (France) **50**, 1689 (1989); M. de Boissieu, C. Janot, J. M. Dubois, M. Audier, M. V. Jarić and B. Dubost, Proc. Anniversary Adriatico Research Conference on Quasicrystals, ed. M. V. Jarić and S. Lundquist (World Scientific, Singapore, at press).

Ideal Structures of the Icosahedral Al-Mn and Al-Cu-Li Quasicrystals

A. Yamamoto

National Institute for Research in Inorganic Materials,
Namiki 1, Tsukuba 305, Japan

Abstract. Ideal structures for the two icosahedral phases have been proposed. These are expressed as six-dimensional icosahedral crystals, the real atom arrangement of which is given by the section on an appropriate three-dimensional hyperplane. Both models have dense packing of icosahedral atom clusters and linking atoms joining them and give reasonable chemical composition and diffraction intensity. Furthermore the model for the Al-Mn or Al-Cu-Li leads to an ideal (b.c.c.) α-Al-Mn-Si or R-Al-Cu-Li structure and an infinite number of other simple cubic crystals when appropriate phason strain is taken into account. A fictitious cubic structure with $a=a_0\tau$ is derived from the model for each case ($\tau=(1+\sqrt{5})/2$, a_0 being the lattice constant of α-Al-Mn-Si or R-Al-Cu-Li). It has seven 12-fold vertices in a unit cell, each of which is occupied by the icosahedral cluster. The simple cubic α-Al-Mn-Si structure is derived from a face-centered icosahedral crystal, which is a superstructure with the lattice constant twice as large as the previous one.

1. Introduction

Recently, the progress in the structure analysis of quasicrystals gave the concrete image of the icosahedral Al-Mn and Al-Cu-Li quasicrystals. We proposed a model of Al-Mn based on x-ray powder diffraction, which consists of many atom clusters called Mackay icosahedra and linking atoms joining them [1]. The Mackay icosahedra are placed at most 12-fold vertices of the three-dimensional Penrose pattern (3DPP), which are joined with an acute rhombohedron or a rhombic dodecahedron. Elswijk et al. gave a model for Al-Cu-Li based on single-crystal x-ray diffraction [2]. This suggests that Al-Cu-Li is also composed of many atom clusters placed at the same 12-fold vertices and linking atoms but the cluster is different from the Mackay icosahedron.

The characteristic feature of the icosahedral quasicrystals is considered to be the existence of such atom clusters at the 12-fold vertices and linking atoms joining them as in their crystal approximants, α-Al-Mn-Si and R-Al-Cu-Li [2-6]. Once we obtain a model for the quasicrystal a related crystal structure can be derived by taking into account appropriate phason strain [3,7,8]. It is known that the α phase in Al-Mn-Si and the R phase in Al-Cu-Li are good crystal approximants for the Al-Mn and Al-Cu-Li quasicrystals: their diffraction intensity distribution is quite similar to that of the quasicrystals [2,3]. Therefore it is interesting to note whether the models lead to such structures when the phason strain is considered. Both models do not however give the real crystal structures.

In the present paper, we consider ideal models which lead to the crystal approximants mentioned above. However such a quasicrystal is not unique because an infinite number of models with such an atom arrangement are possible. Another criterion is necessary to fix the structure. The following guiding principle is employed in this paper. The crystal approximant consists of two units in both cases, that is, the acute rhombohedron and the rhombic dodecahedron, which link two 12-fold vertices [3]. The difference in α-Al-Mn-Si and R-Al-Cu-Li exists only in the decoration of them. Therefore we consider such structure units whenever the 12-fold vertices are linked with the acute rhombohedron and the rhombic icosahedron in the 3DPP. This reduces the ambiguity in the linking atom arrangement of the quasicrystal.

The quasicrystal structure derived in this way will lead to other fictitious crystal approximants with a larger period, in which the same atom clusters appear at the 12-fold vertices and they are linked with the same units as in the α phase or the R phase. In fact it is possible to derive such fictitious crystals with the lattice constant τ times larger than that of the α or R phase. They are available to investigate properties of quasicrystals by means of any conventional theory for usual crystals. The aims of the present work are to demonstrate the construction of ideal quasicrystal structures with such a nature and to show the structures of the fictitious crystal approximants.

2. Simple models

First we consider a simple model for each case in order to show the method used in building the occupation domain. The section method is quite convenient to describe the quasicrystal structures [1]. In this method the quasicrystal structure is represented in the six-dimensional space as a crystal with some occupation domains. It is well known that the 3DPP is obtained from an icosahedral crystal with the triacontahedral occupation domain with unit edge length. (See Fig. 1.) This structure is a basic structure in considering real quasicrystals. It consists of two rhombohedra, called the acute and obtuse ones, all the faces of which are composed of the same (golden) rhombus. Real quasicrystal structures are derived from the 3DPP by the decoration. The way of decoration is different between Al-Mn and Al-Cu-Li.

Simple decorations of the 3DPP have been shown by Ishihara & Shingu [9]. According to them, all the edge centers of the 3DPP are derived from the rhombic icosahedron with unit edge length at the edge center, (100000)/2, in the icosahedral unit cell (and its equivalent positions related with it by symmetry operations of the icosahedral group). (We use the icosahedral coordinate system used in [1].) Similarly, they showed that the face-center positions are obtained from the rhombic dodecahedron at (110000)/2 and the body-center positions of the acute rhombohedron from the acute rhombohedron at (111000)/2. (Fig. 1(b-d).) In fact, in Al-Cu-Li, the vertex and edge-center positions are occupied by Al/Cu except for exceptional cases which will be mentioned later.

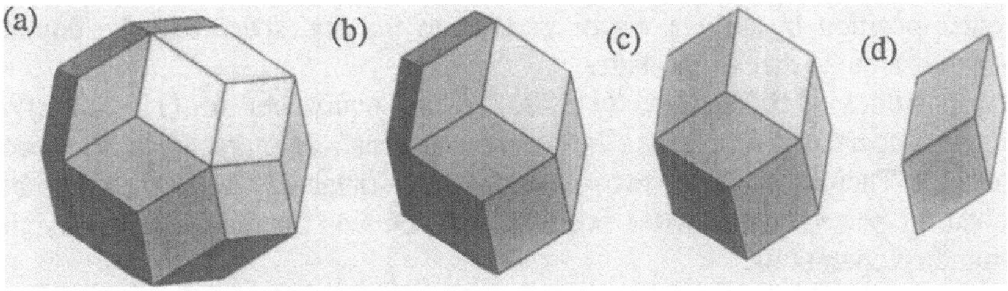

Fig 1 Occupation domains appearing in the three-dimensional Penrose pattern and its simple decoration. (a) rhombic triacontahedron, (b) rhombic icosahedron, (c) rhombic dodecahedron and (d) acute rhombohedron with unit edge length.

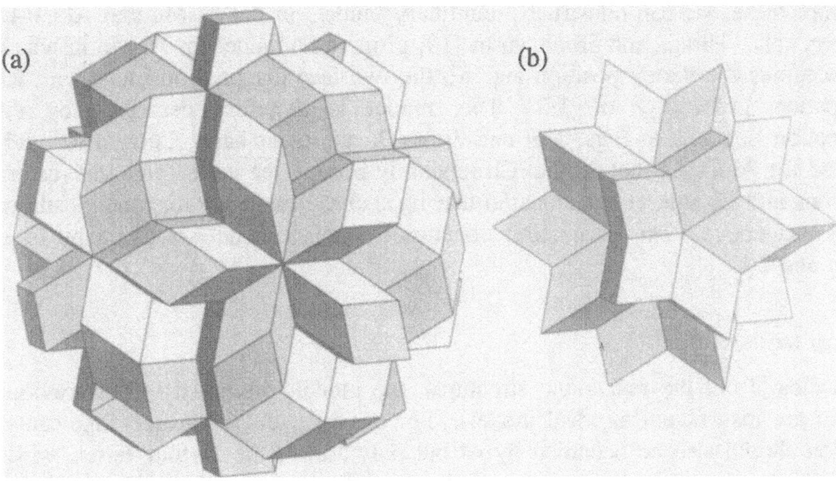

Fig. 2. Occupation domains for (a) two face-diagonal positions and (b) two body-diagonal positions.

We necessitate other positions to describe the real quasicrystal structures, which are the face diagonal two positions and body-diagonal two positions of the acute rhombohedron in the 3DPP. These divide the face and body diagonals into $\tau^{-2} : \tau^{-3} : \tau^{-2}$ ($\tau=(1+\sqrt{5})/2$). They are obtained from the occupation domains at the origin, (000000), and at the body center, (111111)/2 (Fig. 2). Figure 2(a) is composed of the 30 rhombic dodecahedra placed on the unit rhombic triacontahedron (Fig. 1(a)). On the other hand, Fig. 2(b) is the flower-shaped dodecahedron at the body center which is constructed with the 20 acute rhombohedra.

The former domain is derived from the following consideration. As shown by Ishihara and Shingu, the face-center position is obtained from the rhombic dodecahedron placed at (110000)/2. The face diagonal positions are shifted from the face-center position by $\pm\tau^{-3}(110000)^{e}/2$ along the real (external or parallel) space, where the superscript e means the real space component of the six-dimensional vector. It is clear that these two positions appear on all the rhombic faces since the position of the occupation domain is different from that of the face-center position by only a vector parallel to the real space and the domain shape is the same as that of the latter.

The positions $(110000)/2 \pm \tau^{-3}(110000)^e/2$ are equivalent to $(11\text{–}200\text{–}2)^i/2$, where the superscript i means the pseudo (internal or perpendicular) space component. Therefore these two positions are obtained from the rhombic dodecahedron placed at the latter position and domains obtained from it by the site-symmetry operations.

Similarly, the body-diagonal two positions are obtained from the acute rhombohedra placed at $(111000)/2 \pm \tau^{-3}(111000)^e/2$, which are equivalent to $(111111)/2+(000\text{–}11\text{–}1)^i/2$. The domains obtained from it by the site-symmetry operations construct the flower-shaped dodecahedron at the body center.

From these, we can construct preliminary models in the Al-Mn and Al-Cu-Li quasicrystals. Hiraga and Hirabayashi [10] proposed a model for Al-Mn in which Mn occupies the vertex position and Al, the two face diagonal positions with the occupation probability of 1/2. This model is therefore described by the occupation domains in Figs. 1(a) and 2(a). On the other hand, Chen et al. [11] showed for Al-Cu-Li that Al and Cu randomly occupy the vertex and edge-center positions and Elswijk et al. indicated that Li is at the two body-diagonal positions mentioned above. The occupation domains of such a model are given by Figs. 1(a,b) and 2(b).

3. Ideal models

As is clear from the real cubic structures, the models discussed in the previous section are insufficient as ideal models. For example, in Al-Mn, the edge-center position should also be occupied by Al but only around the 12-fold vertex, while most Al atoms are situated at one of the face diagonal positions. In Al-Cu-Li, some edge-center Al/Cu should be slightly shifted. This new position divides the edge into $\tau^{-1} : \tau^{-2}$.

(a) Al-Mn. In the ideal model of Al-Mn, we place an atom cluster, the so-called Mackay icosahedron, at each 12-fold vertex, in which 12 Al atoms are at $(100000)^e/2$ from the center (and its equivalent positions), 12 Mn at $(100000)^e$ and 30 Al at $\tau^{-1}(110000)^e$. The 12-fold vertices are linked with the rhombic dodecahedron or the acute rhombohedron in the 3DPP but small amount of them are at the positions with nearest neighbor distance of unit length [12]. Then the Mackay icosahedra are interpenetrated. In such a case, we place the Mackay icosahedron at only one of the two vertices.

The 12-fold vertices occupied by the Mackay icosahedron are given by the occupation domain shown in Fig. 3(a). This is obtained from the small triacontahedron with the edge length of τ^{-2} by truncating tips along the five-fold axes. Hereafter we call such 12-fold vertices the 12-fold vertices for simplicity. The inner 12 Al, 12 Mn and outer 30 Al in the Mackay icosahedron are called Al(α), Mn(α) and Al(β) atoms in order to distinguish them from other Al or Mn [1].

The Al(α) position is obtained from two domains (Fig. 3(b)) placed at $(100000)/2 \pm (100000)^i/2$. Here the small truncated triacontahedra share ten faces around the 5-fold axis. (These domains are included in the domain for the edge-

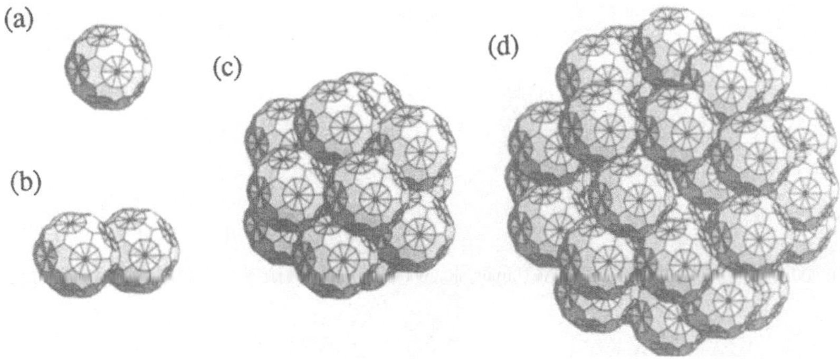

Fig. 3. Occupation domains for the Mackay icosahedron. (a) the small truncated triacontahedron for the 12-fold vertex, (b) domains for Al(α), and those of (c) Mn(α) and (d) Al(β).

center position shown in Fig. 1(b) because the Al(α) atoms are a part of the edge-center atoms.) The Mn(α) atoms are derived from the occupation domains shown in Fig. 3(c) and Al(β) atoms, from Fig. 3(d) [1]. For all the Mn atoms we consider the occupation domain shown in Fig. 4(a) placed at the origin. This is obtained from the unit triacontahedron by subtracting the Al(β) domain. (The figure is drawn by subtracting small triacontahedra instead of the small truncated triacontahedra as in Fig. 3(a) for simplicity.) In the α phase, the 12-fold vertex is vacant. We assume this in the quasicrystal. Then the shape of Fig. 2(a) must be removed from the central part of Fig. 4(a). This domain includes the Mn(α) domain.

For Al atoms at one of the face diagonal positions mentioned in the previous section, we can consider the domain shown in Fig. 4(b). This has a vacant core with the shape of Fig. 1(a) and is obtained from Fig. 2(a) by cutting the outer half of the rhombic dodecahedron. As shown by us [1] and Duneau & Oguey [13], this occupation domain should be modified by cutting the tips on the 5-fold axes and removing the common part with the small truncated triacontahedra placed at $(111000)^e$ and its equivalent positions in order to prevent the appearance of the short Al-Al distances. (See Fig. 4(c).) The domain includes only a part of the Al(β) domain. Therefore the total Al domain is the union of these two.

Other Al atoms are present in the α phase: when the two 12-fold vertices are linked with the rhombic dodecahedron, the body-diagonal positions of the two acute rhombohedra in the rhombic dodecahedron are occupied by Al (Al(δ)) [3]. Therefore when two 12-fold vertices are linked with the rhombic dodecahedron in the 3DPP, we consider the Al(δ) at the same place in the dodecahedron. The occupation domain for this atom (Fig. 4(d)) is placed at the body center (111111)/2. This is derived from the common part of the small triacontahedra placed at the origin and $(111100)^i$ by a similar consideration. The domain has a vacant core with the shape of Fig. 4(e) at the center. (Exactly speaking, the small triacontahedron should be replaced by the small truncated triacontahedron in the above and following description but because of the difficulty in the graphical representation we use the former.)

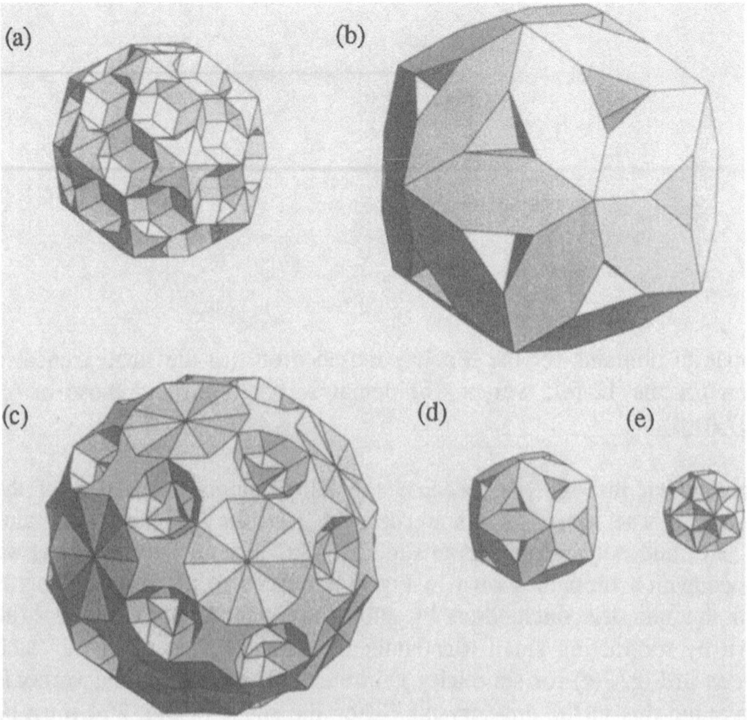

Fig. 4. Occupation domains for (a) Mn, (b) one of the face-diagonal positions, and (c) Al in Al-Mn. Occupation domains for (d) Al(δ) and (e) its vacant part at the center.

(b) Al-Cu-Li. As shown by Elswijk et al., the model described in the previous section gives a good approximation for the real quasicrystal, in which the Al/Cu atoms occupy all the vertex and edge-center positions and Li atoms, all the body-diagonal positions. In order to obtain the R phase as a crystal approximant, it must however be modified as follows.

We assume an atom cluster at every 12-fold vertex, which is composed of 12 Al/Cu atoms at $(100000)^e/2$ around the center and other 12 Al/Cu atoms at $(100000)^e$. These are the same as the Al(α) and Mn(α) positions in Al-Mn. On the other hand, 20 Li atoms are situated at $\tau^{-2}(111000)^e$. These are called Al/Cu(α), Al/Cu(β) and Li(α) positions. Consequently the atom cluster in Al-Cu-Li consists of 24 Al/Cu and 20 Li atoms. It is clear that the occupation domains of Al/Cu(α) and Al/Cu(β) are the same as those of Al(α) and Mn(α) in Al-Mn and given by Figs. 3(b,c). For the Li(α) position, we can consider the occupation domain consisting of small truncated triacontahedra placed at $\tau^{-2}(111000)^e$ (and its equivalent positions). This is equivalent to putting the same triacontahedra at $(111111)/2-(111-11-1)^i/2$. These domains correspond to the small triacontahedra in Fig. 5(b).

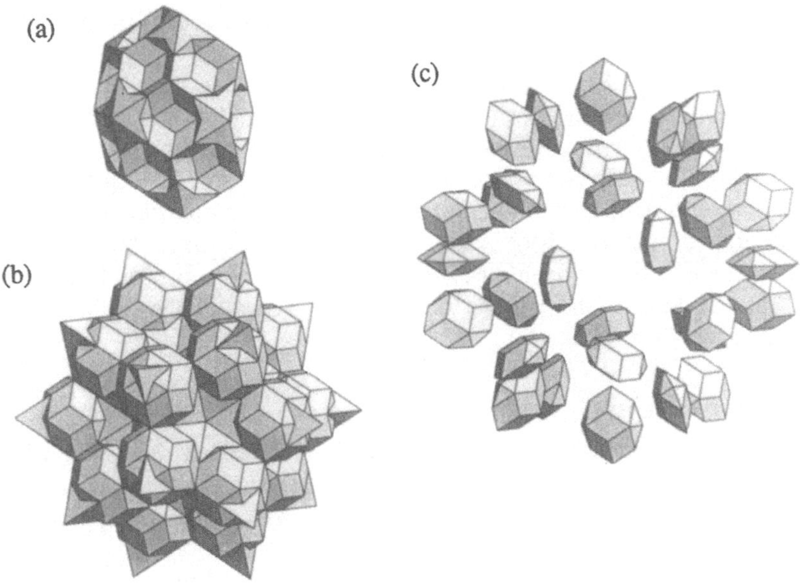

Fig. 5. Occupation domains for (a) the edge-center position of Al/Cu (b) Li and (c) the edge-off-center position of Al/Cu.

When the atom cluster mentioned above is considered at each 12-fold vertex some atoms in the previous model must be removed in order to prevent the appearance of very short interatomic distances. For example if we consider the unit triacontahedron for the vertex position, this sometimes causes an atom at the position of Al(β) in Al-Mn because the domain of the vertex position and that of Al(β) are partly overlapped. In Al-Cu-Li, there is no atom at the position of Al(β) in contrast to Al-Mn. Therefore we have to cut the common part with the Al(β) domain for the Al/Cu domain. As a result the occupation domain agrees with that of Mn in Al-Mn, but in this case the central part of this domain is not empty because the 12-fold vertex is considered to be (partly) occupied by Al/Cu. (According to the recent analysis the 12-fold vertex may be empty [14]. Then the domain completely agrees with the Mn domain.)

Similarly, several atoms in the edge-center position have to be removed because some atoms in the rhombic dodecahedron are obtained from the edge-center atoms on shifting along the edge by $\tau^{-1}-1/2=0.1180$. This can be achieved by subtracting a part of the small triacontahedron from each golden rhombus of the rhombic dodecahedron. Then we have a domain shown in Fig. 5(a). The domain for Li atoms (Fig. 5(b)) is a union of the 20 small triacontahedra at $(111111)/2-(111-11-1)^{i}$ and its equivalent positions and the flower-shaped dodecahedron. The parts removed from the rhombic icosahedron are shifted to the convex part on each golden rhombus of the flower-shaped dodecahedron in the figure and 30 small domains, like small rhombic dodecahedra, around the body center (111111)/2 (Fig. 5(c)). Figure 5(c) creates

Al/Cu atom positions (denoted by G in [4]) in the rhombic dodecahedron which links the 12-fold vertices.

In conclusion, in order to obtain the local atom arrangement in the real crystals, we have to consider complicated occupation domains: the related domains are given by Figs. 3(b) and 4(a,c,d) for Al-Mn and by Figs. 4(a) and 5(a-c) for Al-Cu-Li. It is noted that in both models, atoms are located at the vertex position of the 3DPP or the point apart from the vertex position by $\pm 1/2$ or $\pm\tau^{-1}$ along one of the six 5-fold axes. As seen above, these positions are described by the occupation domains at the origin, the edge center and the body center of the six-dimensional icosahedral unit cell. Thus all the positions are described by these three cases.

4. Crystal Approximants

We consider the cubic crystal structures which are derived from the ideal quasicrystal structures mentioned above. It can be shown that the real crystal structure is obtained from the quasicrystal structures by considering appropriate phason strain.

The phason strain is expressed by the shear strain between the real and pseudo spaces. Then appropriate strain leads to a periodic crystal. We can obtain an infinite number of crystal structures, but here we only consider four cubic crystals, which are the ideal crystal structures corresponding to α-Al-Mn-Si and R-Al-Cu-Li ((1,1) structures) and fictitious ones ((2,1) structures) with a larger unit cell dimension by a factor of τ.

(a) (1,1) structures. These are obtained from the models given in the previous section. As shown in Table 1, the real crystal structures are really obtained from the models under appropriate phason strain. Slight modification is necessary for the α-Al-Mn-Si structure because the model leads to a body-centered cubic structure while the real α-Al-Mn-Si is simple cubic. The simple cubic structure is obtained from the above model by removing 6 Al(5) atoms belonging to the one sublattice of the body-centered cubic lattice. Such a structure is obtained from a superstructure in the six-dimensional space. The six-dimensional primitive lattice is divided into two sublattices consisting of lattice points with even and odd parities. Then each sublattice has the face-centered icosahedral symmetry. If we place the occupation domain of Al(δ) only on one sublattice, we will obtain the simple cubic structure. The atom positions obtained from the models are listed in Table 1 together with those of the (2,1) structures.

(b) (2,1) structures. As shown in the table, the Al-Mn (2,1) structure includes 409 Al and 98 Mn atoms while Al-Cu-Li has 447 Al/Cu and 211 Li. Their chemical compositions, $Al_{80.7}Mn_{19.3}$ and $(Al,Cu)_{67.9}Li_{32.1}$, are very close to those of the quasicrystals. The structures are very complicated but their framework, which is given by the vertex model, is comparatively simple. The model consists of the flower-shaped dodecahedron at the origin, the rhombic triacontahedron at the body center and 9 rhombic dodecahedra which are linked with 34 acute and 24 obtuse rhombohedra. It is noted that this structure includes seven 12-fold vertices, the positions of which agree with those of Al/Cu(1) and Al/Cu(5). The

Table 1. The structural parameters of ideal cubic Al-Mn and Al-Cu-Li structures. The space groups of the (1,1) and (2,1) structures are Im3 and Pm3. (e=0.19098, f=1/2–e=0.30902, g=0.11803, h=1/2–g=0.38197, i=0.09549, j=0.15451, k=1/2–j=0.34549, l=0.42705, m=0.07295, n=0.05902, p=0.23607, q=1/2–p=0.26393, r=1/2–i=0.40451, s=1/2–n=0.44098, t=0.28648, u=0.13197, v=1/2–t=0.21352, w=1/2–u=0.36803, x=0.45492, y=0.35410)

Al-Mn (1,1)					Al-Cu-Li (1,1)				
atom		x	y	z	atom		x	y	z
Mn	24g	f	0	e	Al/Cu(1)	2a	0	0	0
Al(1)	24g	0	i	j	Al/Cu(2)	24g	f	0	e
Al(2)	48h	g	f	e	Al/Cu(3)	24g	0	i	j
Al(3)	12d	h	0	0	Al/Cu(4)	48h	f	i	k
Al(4)	24g	g	g	1/2	Al/Cu(5)	12e	0	1/2	l
Al(5)	12e	0	1/2	e	Li(1)	12e	0	1/2	e
					Li(2)	16f	e	e	e
					Li(3)	24g	0	f	g
Al-Mn (2,1)					Al-Cu-Li (2,1)				
Mn(1)	12j	e	0	g	Al/Cu(1)	1a	0	0	0
Mn(2)	6e	0	h	0	Al/Cu(2)	12j	e	0	g
Mn(3)	24l	e	g	f	Al/Cu(3)	6e	0	h	0
Mn(4)	8i	f	f	f	Al/Cu(4)	24l	e	g	f
Mn(5)	12j	h	h	0	Al/Cu(5)	6g	0	1/2	e
Mn(6)	12k	e	1/2	f	Al/Cu(6)	8i	f	f	f
Mn(7)	12k	m	e	1/2	Al/Cu(7)	12j	h	h	0
Mn(8)	12k	h	1/2	e	Al/Cu(8)	12k	e	1/2	f
Al(1)	12j	0	n	i	Al/Cu(9)	12k	m	e	1/2
Al(2)	24l	m	e	g	Al/Cu(10)	12k	h	1/2	e
Al(3)	6e	p	0	0	Al/Cu(11)	12j	0	n	i
Al(4)	12j	h	0	p	Al/Cu(12)	24l	e	n	v
Al(5)	12j	q	e	0	Al/Cu(13)	24l	1/4	i	g
Al(6)	24l	m	f	f	Al/Cu(14)	12j	n	t	0
Al(7)	24l	m	m	f	Al/Cu(15)	12j	0	s	i
Al(8)	24l	n	r	e	Al/Cu(16)	24l	1/4	v	f
Al(9)	12j	0	s	t	Al/Cu(17)	24l	n	r	e
Al(10)	12k	i	1/2	u	Al/Cu(18)	12j	s	t	0
Al(11)	12j	0	s	i	Al/Cu(19)	12k	1/2	1/4	i
Al(12)	12k	i	1/2	1/4	Al/Cu(20)	24l	f	w	r
Al(13)	12j	l	0	g	Al/Cu(21)	24l	1/4	r	f
Al(14)	24l	p	g	e	Al/Cu(22)	12k	u	i	1/2

Table 1. (continued)

Al-Mn (2,1)					Al-Cu-Li (2,1)				
atom		x	y	z	atom		x	y	z
Al(15)	24l	h	g	l	Al/Cu(23)	24l	h	s	i
Al(16)	12k	1/2	e	p	Al/Cu(24)	24l	s	r	e
Al(17)	24l	q	f	e	Al/Cu(25)	12k	1/4	t	1/2
Al(18)	12k	m	l	1/2	Al/Cu(26)	24l	f	g	p
Al(19)	24l	h	h	p	Al/Cu(27)	24l	h	m	q
Al(20)	24l	l	g	f	Al/Cu(28)	6g	x	1/2	0
Al(21)	24l	q	e	h	Al/Cu(29)	6h	1/2	1/2	y
Al(22)	12k	h	1/2	l	Al/Cu(30)	6h	1/2	1/2	q
Al(23)	12k	l	p	1/2	Li(1)	12j	0	f	g
Al(24)	12j	e	f	0	Li(2)	24l	g	h	m
Al(25)	6g	f	1/2	0	Li(3)	8i	g	g	g
Al(26)	1b	1/2	1/2	1/2	Li(4)	12j	e	f	0
					Li(5)	12j	0	e	m
					Li(6)	6g	m	1/2	0
					Li(7)	24l	h	f	g
					Li(8)	6g	f	1/2	0
					Li(9)	24l	l	e	e
					Li(10)	12k	1/2	h	m
					Li(11)	8i	e	e	e
					Li(12)	12j	0	f	q
					Li(13)	6h	1/2	1/2	g
					Li(14)	24l	h	e	f
					Li(15)	8i	l	l	l
					Li(16)	12k	h	f	1/2
					Li(17)	1b	1/2	1/2	1/2

characteristic feature of the structure is the appearance of the rhombic triacontahedron and the flower-shaped dodecahedron as in the 3DPP but the rhombic icosahedron is still absent.

5. Concluding remarks

The ideal icosahedral quasicrystal structures are given for the icosahedral Al-Mn and Al-Cu-Li quasicrystals. The structures are constructed so that their local atom arrangement is similar to that of the real crystal approximants and they lead to the ideal structure of α-Al-Mn-Si and R-Al-Cu-Li when appropriate phason strain is taken into account. Their occupation domains are represented by complicated polyhedra with convex and concave planes. The fictitious cubic (2,1) structures are derived from the models.

Acknowledgment

The author thanks Dr. K. Kato, National Institute for Research in Inorganic Materials, for the help of graphical representation of the occupation domains.

References

1. A. Yamamoto and H. Hiraga, Phys. Rev. B **37**, 6207 (1988).
2. H. B. Elswijk, J. Th. M. de Hosson, S. van Smaalen, and J. L. de Boer, Phys. Rev. B **38**, 1681 (1988).
3. V. Elser and C. L. Henley, Phys. Lett. **55**, 2883 (1985).
4. C. L. Henley, Phil. Mag. B **53**, L59 (1986).
5. M. Cooper and K. Robinson, Acta Crystallogr. **20**, 614 (1966)
6. E. E. Cherkashin, P. I. Kripyakevich, and G. I. Oleksiv, Cristallographiya**8**, 681 (1964).
7. K. Kramer, Acta Cryst. A **43**, 486 (1987).
8. Y. Ishii, Phys. Rev. B **39**, 11862 (1989).
9. K. N. Ishihara and P. H. Shingu, J. Phys. Soc. Jpn. **55**, 1795 (1986).
10. K. Hiraga and M. Hirabayashi, in Proceedings of the First International Symposium for Science of Form, University of Tsukuba, 1985 (unpublished).
11. Y. Chen, S. J. Poon, W. Dumowski, T. Egami, and G. J. Shiflet, Phys. Rev. Lett. **58**, 1440 (1987).
12. C. L. Henley, Phys. Rev. B **34**, 797 (1986).
13. M. Duneau and C. Oguey, J. Phys. France **50** 135 (1989).
14. M. V. Jarić, private communication, (1989).

High-Resolution Electron Microscopy and Atomic Arrangement of an Al-Li-Cu Quasicrystal

K. Hiraga

Institute for Materials Research, Tohoku University, Katahira, Aoba, Sendai 980, Japan

An icosahedral quasicrystal of Al-Li-Cu was studied with a 400 kV high-resolution electron microscope having a resolution of 0.17 nm. I succeeded in observing high-resolution structure images which faithfully represent the projected potential of the quasicrystal. The observed images were interpreted well by an atom cluster model based on the three-dimensional Penrose tiling. In the model, atom clusters forming rhombic triacontahedra are placed at twelvefold sites in the three-dimensional Penrose tiling. General aspects of high-resolution electron microscopy of quasicrystals are discussed.

1. Introduction

High-resolution electron microscopy of quasicrystals gave us valuable information about structures of quasicrystalline alloys; e.g. it ruled out the models which were proposed to explain the diffraction patterns with icosahedral symmetry in terms of a large unit cell or multiply twinned crystals, and showed the existence of fivefold symmetry and long-range quasiperiodicity in real space [1]. Also high-resolution observations revealed interesting defects in the quasicrystals, such as quenched phason strains [2] and dislocations [3]. However, almost all the images observed so far are so-called lattice images taken with 200 kV electron microscopes having a resolution of about 0.23 nm and so have given us little information about atomic arrangement of the quasicrystals.

The purpose of this work is to present so-called high-resolution structure images taken from thin regions where multiple scattering is lowered, by using a 400 kV electron microscope having a resolution of 0.17 nm, and is to discuss atomic arrangement of the quasicrystal. In this paper, we will deal with a stable Al-Li-Cu quasicrystal which is less affected by multiple scattering than the other quasicrystalline alloys, because of a small average atomic number.

An outline of the rest of the paper is as follows. In Section 2, I discuss general aspects of high-resolution electron microscopy of quasicrystals, particularly with attention to the difference between the lattice images and structure images. Section 3 considers an atomic model of the Al-Li-Cu quasicrystal for comparison with observed high-resolution structure images. In the last section, I show a correspondence between the observed images and the structure model.

Springer Series in Solid-State Sciences, Vol. 93 Quasicrystals
Editors: T. Fujiwara · T. Ogawa

2. High-resolution electron microscopy of a quasicrystal

Figure 1 shows two types of high-resolution electron micrographs of an Al-Mn-Si quasicrystal taken with the incident beam parallel to the fivefold symmetry axis, by using a 200 kV and a 1 MV electron microscope, which have a resolution of 0.23 nm and 0.16 nm, respectively. Figure 1 (a), which is a typical high-resolution image presented by many researchers so far, clearly shows topological features of quasilattices; i.e. an array of bright dots on straight lines parallel to the fivefold directions and lattice fringes (bright dot rows) shows quasiperiodicity. However, this image is a so-called lattice image taken from a relatively thick region where intensities of the weak reflections with lattice spacings of longer than 0.23 nm are enhanced by the effect of multiple scattering. This situation can be clearly seen in a Fourier transformed pattern of the image (Fig. 2 (a)). Actually, the diffraction spots appearing in the pattern are very weak in the kinematical approximation and are hardly detected in powder X-ray diffraction. Main peaks which result from neighboring pairs of atoms, having strong intensity in X-ray diffraction, make little contribution to the image.

On the other hand, the image taken with an 1MV electron microscope (Fig. 1 (b)) shows fine contrast compared with Fig. 1 (a). The image was taken from a thin region (less than 10 nm) where the effect of multiple scattering is not strong. In the image, we can see lattice fringes with an interval of about 0.2 nm by oblique viewing. Also the Fourier transformed pattern of the image (Fig. 2 (b)) shows that the peaks with a lattice spacing of about 0.2 nm, indicated by small arrows, have strong intensity and the spots appearing in Fig. 2(a)

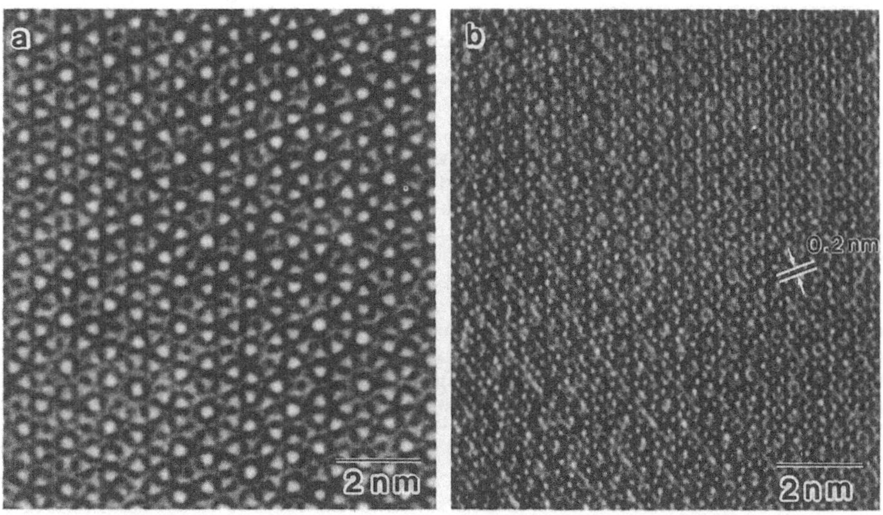

Fig. 1 Two types of high-resolution electron micrographs of an Al-Mn-Si quasicrystal taken with the incident beam parallel to the fivefold axis. (a) and (b) were taken with a 200 kV and an 1 MV electron microscope having a resolution of 0.23 and 0.16 nm, respectively.

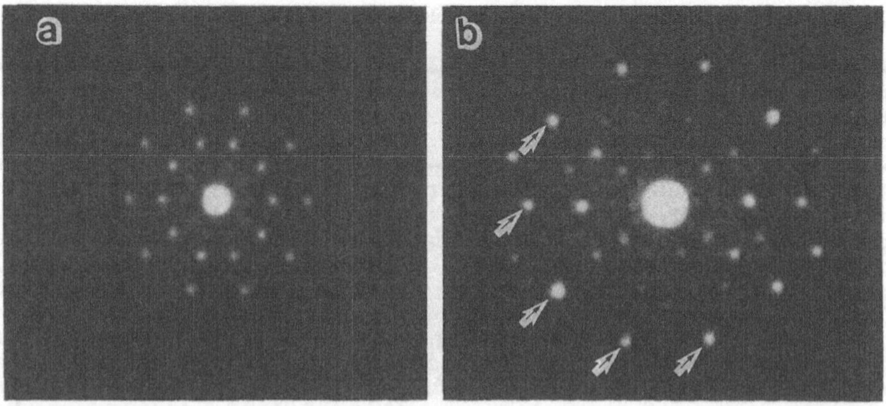

Fig. 2 Fourier transformed patterns of the images of Figs. 1(a) and 1(b). (a) is taken from Fig. 1(a) and (b) is from Fig. 1(b).

become weak. The deviation of their intensity from fivefold symmetry is due to incorrect alignment, i.e. that the incident beam is not exactly parallel to the fivefold axis. This intensity distribution in the Fourier diffractogram is considered to approximately reproduce that in the kinematical approximation. The result shows that the image taken with an 1 MV electron microscope is a so-called structure image which faithfully represents the projected potential of the quasicrystal, and so the image may have valuable information about the atomic arrangement of the quasicrystal. Actually, characteristic image contrasts consisting of a bright ring surrounded by ten bright dots in the image were found to represent the projection of the Mackay icosahedral atom clusters [4].

3. Structure model

Here I will consider a structure model of the Al-Li-Cu quasicrystal, which is constructed by the combination of an atom cluster model and the three-dimensional Penrose tiling. An atom cluster for interpretation of the atomic arrangement of the Al-Li-Cu quasicrystal has been proposed owing to the similarity of a R-AlLiCu crystalline phase, as shown in Fig. 3 (a) [5,6]. In the atom cluster, two sizes of atoms are closely packed; i.e. large atoms placed on the threefold axes are considered to correspond to Li atoms and small ones at the other positions are Al and Cu atoms. Atom positions in the atom cluster are considered as follows. It is known that a polyhedron (Fig. 3(b)) constructed with twenty acute rhombohedra is placed at the vertices of the three-dimensional Penrose tiling inflated with a scaling parameter of τ^3 . The polyhedron has icosahedral symmetry, and so atom clusters having icosahedral symmetry can be formed by placing atoms at special positions in the twenty acute rhombohedra. An atomic decoration in one of the twenty rhombohedra for the atom cluster of Fig. 3(a) is shown in Fig. 3(c). In the decoration, the successive atomic shells centering the 0 atom form various

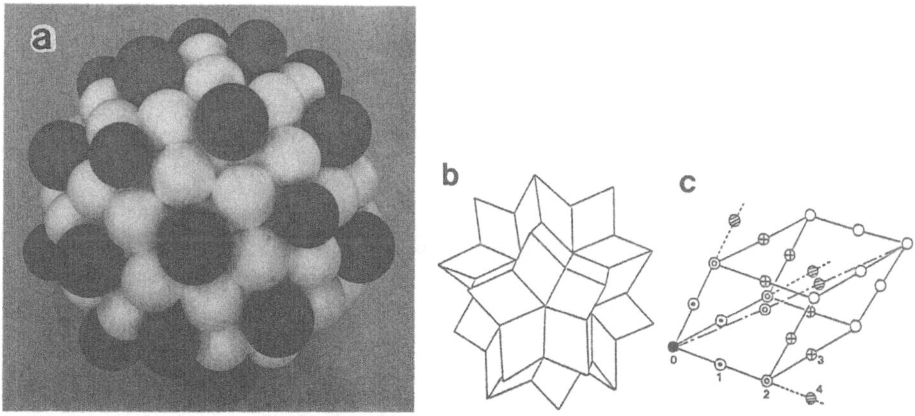

Fig. 3 (a): triacontahedral atom cluster constructed with 137 atoms. (b): polyhedron constructed with twenty acute rhombohedra. (c): atomic decoration in one of the twenty acute rhombohedra. Numbers in (c) indicate the order of successive atomic shells.

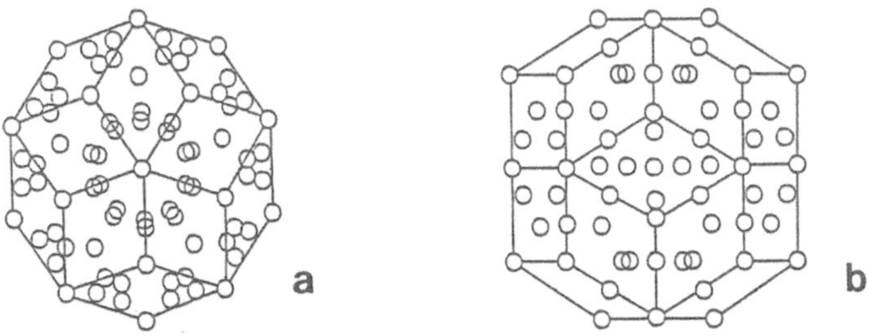

Fig. 4 Atomic arrangements of the triacontahedral atom cluster (Fig. 3(a)), projected along the fivefold (a) and twofold axes (b).

types of polyhedra with an icosahedral symmetry; i.e. icosahedron (first shell), triacontahedron (second shell), truncated icosahedron (third shell) and a large triacontahedron (4th shell) are formed by slightly shifting positions from the ideal ones of Fig. 3(c) [6]. Figures 4(a) and 4(b) show projections of the atom positions in the cluster of Fig. 3(a) along the fivefold and twofold symmetry axes, respectively.

The atom cluster forms a triacontahedron with an edge length of a, which is the same as the edge length of the fundamental rhombohedra forming the three-dimensional Penrose tiling. If we place the triacontahedra at the twelvefold sites in the three-dimensional Penrose tiling [7], the space may be filled up with three types of linkages shown in Fig. 5, although the complete packing of the triacontahedra in space has not been successfully proved at present. The three types of linkages are connected by sharing a rhombic plane (a), an obtuse rhombohedron (b) and an rhombic icosahedron (c). The frequency of the three types of linkages in the three-dimensional Penrose tiling has already been calculated by

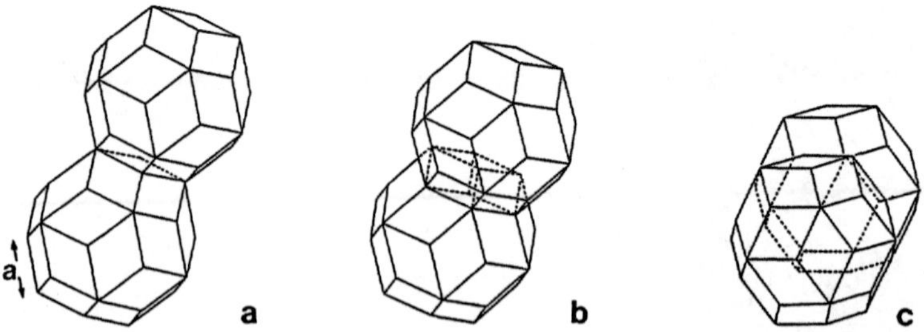

Fig. 5 Three types of linkages of the triacontahedra in the three-dimensional Penrose tiling.

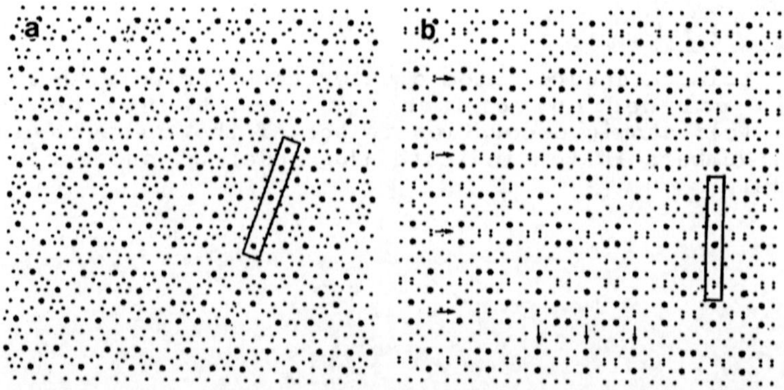

Fig. 6 Twelvefold sites in the three-dimensional Penrose tiling, projected along the fivefold (a) and twofold axes (b). Small dots correspond to vertices of the fundamental rhombohedra and large dots are the twelvefold sites.

Henley [7]; major linkages are the (a) and (b) types and the frequency of the (c) type of linkages is extremely low. Also it was found that the structure of the R-AlLiCu crystalline phase can be considered to be formed by the (a) and (b) types of linkages of the triacontahedron atom clusters. Similar structure models where atom clusters having icosahedral symmetry are placed at the twelvefold sites in the three-dimensional Penrose tiling have already been proposed to interpret the structure of the Al-Mn quasicrystal [8,9].

For comparison with high-resolution images taken with the incident beam parallel to the fivefold and twofold axes, the twelvefold sites projected along the fivefold and twofold axes in the three-dimensional Penrose tiling are shown in Fig. 6.

4. High-resolution structure images and comparison with structure model

High-resolution observations of the Al-Li-Cu quasicrystal were performed with a 400 kV electron microscope (JEM-4000EX) having resolution of 0.17 nm. Figure 7 shows a high-resolution electron micrograph of the Al-Li-Cu quasicrystal taken with the incident beam parallel to the fivefold axis, together with an electron diffraction and a Fourier transformed pattern of the image. The Fourier transformed pattern reproduces well the characteristic features of the electron diffraction pattern, considering that the intensity of weak spots in the electron diffraction pattern is enhanced by the effect of multiple scattering. The thickness of the region of the image is considered to be 5-10 nm, and the image was taken under an optimum defocus of about 35 nm, which was confirmed from the Fresnel fringes observed at the edge of an amorphous film in the image. Thus, the observed image is a so-called structure image and is considered to represent the faithfully projected potential of the quasicrystal. Thus we can say that dark and bright regions in the image correspond to high- and low-potential regions, respectively. In the image, we notice characteristic image contrasts which are composed of a bright ring surrounded by ten bright dots. Hereafter, we call the characteristic image contrast decagon contrast. We can expect the decagon contrast from the projected atomic arrangement of Fig. 4(a); i.e. a ring-shaped channel around the centered atom becomes a bright ring and channels around the ring channel make double decagons of bright dots. Considering that the outer ten bright dots are always overlapped with those in the other decagon contrasts, the image contrast expected from the tiacontahedral atom cluster is consistent with the decagon contrast in the observed image.

In the image of Fig. 7, the decagon contasts are apparently arrayed on straight lines parallel to the fivefold directions with quasiperiodicity, but close examination over the wide region with oblique viewing shows a few steps of the alignments, which result from quenched phason strains. In the Al-Li-Cu quasicrystal, I have always observed the phason strains and thus we have to take account of their existence when we discuss the ideal structure of the quasicrystal. From the high-resolution observations, I can say that the decagon contrasts are arrayed on straight lines along the fivefold directions in the ideal quasicrystal. In an enlarged micrograph of a part of Fig. 7 (Fig. 8(a)), we can clearly see an arrangement of the decagon contrasts with some distances related to the golden ratio τ along the fivefold directions, as indicated with arrows. Figures 8(b) and 8(c) show an atomic arrangement and expected image contrast for the case that the triacontahedron atom clusters are placed at the twelvefold sites in the rectangle of Fig. 6(a). The expected contrast (c) shows a good correspondence with the observed image contrast; the decagon contrasts are arrayed with three types of distances related to τ , and some bright rings are divided into three or two parts by overlapping of the decagon contrasts. So we can conclude that the observed image contrast can be interpreted well by the model mentioned in the previous section. The conclusion is also confirmed from an observed image taken with the incident beam parallel to the twofold axis.

Figure 9 is a high-resolution image taken with the incident beam parallel to the twofold axis, together with an electron diffraction and a Fourier transformed pattern of the image. Here we should notice in the image that there is a rather high density of phason strains, which bring about frequent displacements of the lattice fringes along the two directions indicated by large arrows. In the image, characteristic image

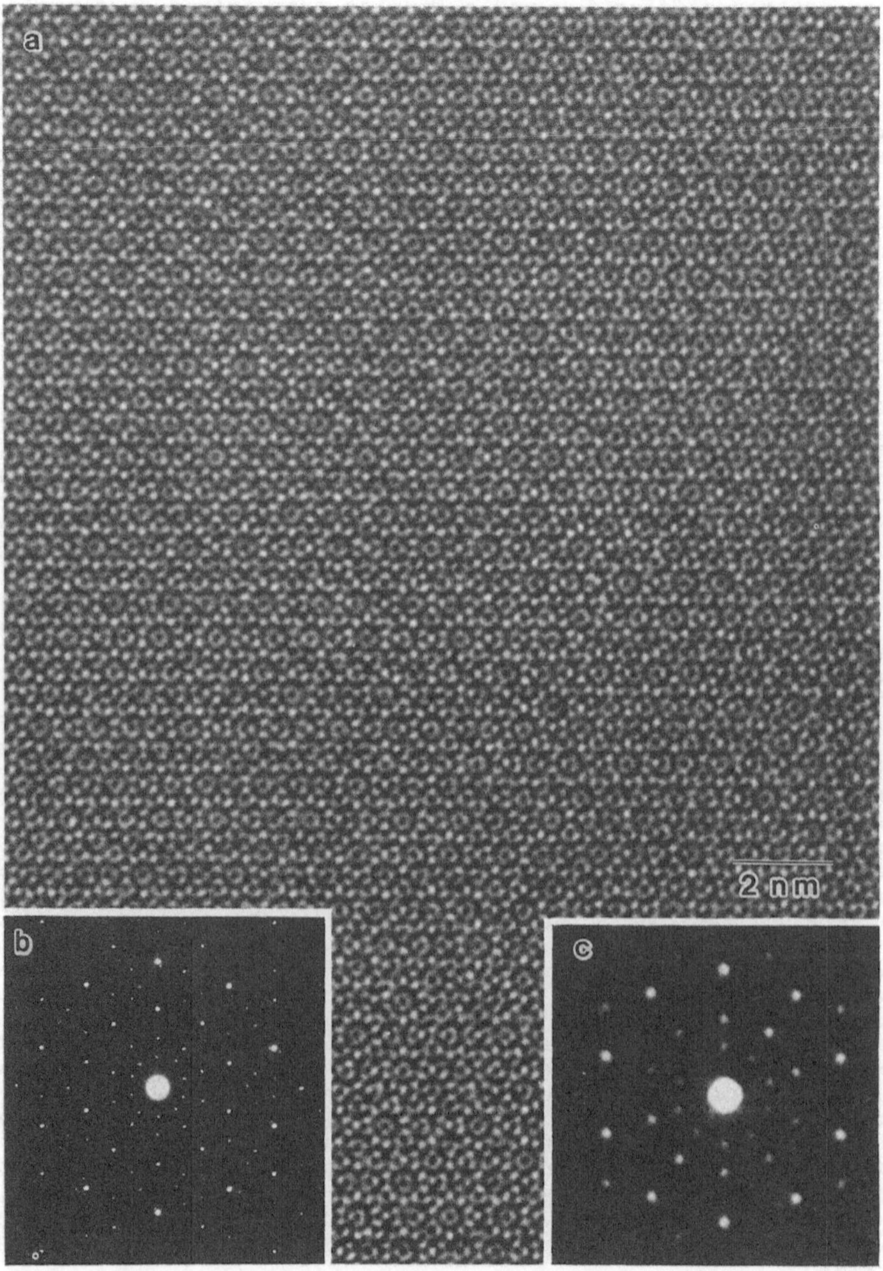

Fig. 7 High-resolution electron micrograph (a) of the Al-Li-Cu quasicrystal taken with the incident beam parallel to the fivefold axis, together with an electron diffraction (b) and a Fourier transformed pattern (c) of the image.

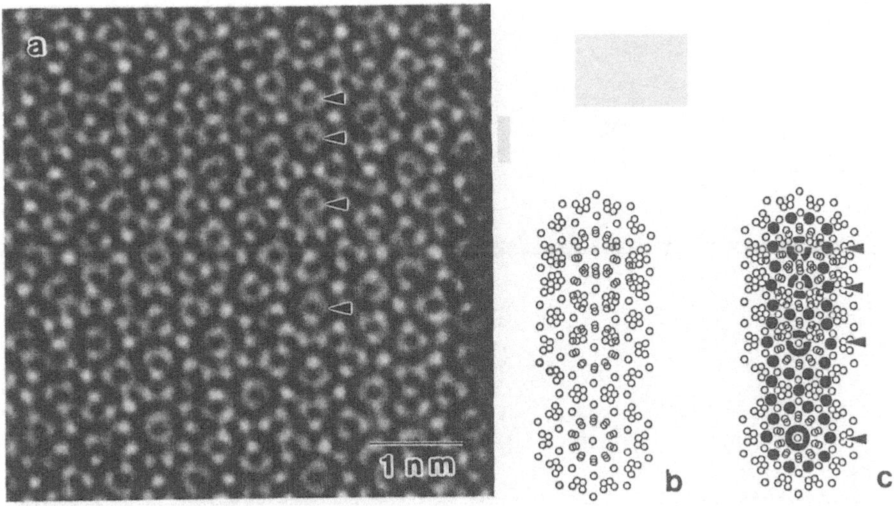

Fig. 8 (a): enlarged picture of a part of Fig. 7(a). (b): atomic arrangement in the rectangle of Fig. 6(a); i.e. the atomic arrangement of Fig. 4(a) is placed at the twelvefold sites. (c): expected image contrast of the atomic arrangement of (b). Black regions correspond to bright regions in the observed image of (a).

contrasts with nearly periodical arrangements which are sometimes interrupted to form incommensurate periodicity, can be seen along the two twofold directions, as indicated with small arrows. Those partial periodicities can be seen in the arrangement of the twelvefold sites, as indicated with arrows in Fig. 6(b).The periodicities with long distances cause the appearance of the diffraction spots indicated by arrows in Fig. 9(b).

Detailed image contrast can be clearly seen in an enlarged picture of a part of Fig. 9(a) (Fig. 10(a)). Also an atomic arrangement and expected image contrast for the case that the atomic arrangement of Fig. 4(b) is placed at the twelvefold sites in the rectangle in Fig. 6(b) are shown in Figs. 10(b) and 10(c), respectively. The expected image contrast (c) shows a good correspondence to the observed image contrast between the small arrows along the vertical direction in Fig. 10(a).

In conclusion, high-resolution microscopy with a 400 kV electron microscope having a resolution of 0.17 nm made it possible to observe structure images which faithfully reproduce the projected potential of the quasicrystal. The observed images were interpreted well by the structure model that the triacontahedral atom clusters are placed at the twelvefold sites in the three-dimensional Penrose tiling. In the model, the triacontahedral atom clusters are densely packed in space with three types of linkages. The high-resolution observations gave us valuable information about the aggregation of the atom clusters in the quasicrystal, but accurate structure analysis about atom positions and atomic ordering in the clusters requires quantitative X-ray and neutron-diffraction experiments.

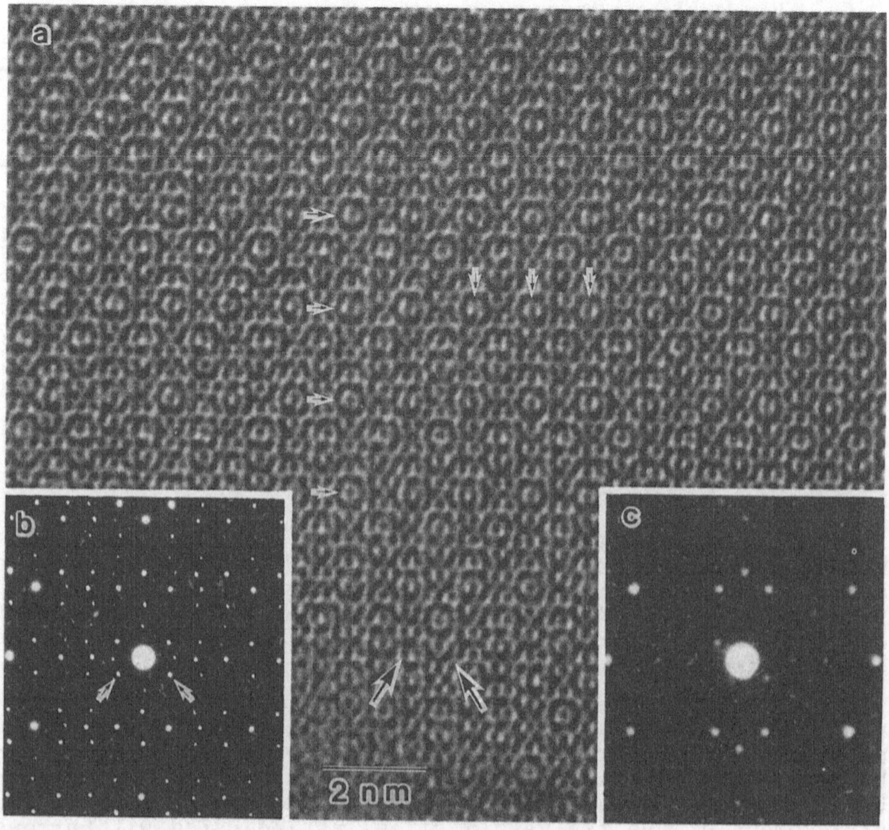

Fig. 9 High-resolution electron micrograph (a) of the Al-Li-Cu quasicrystal, taken with the incident beam parallel to the twofold axis, together with an electron diffraction (b) and a Fourier transformed pattern (c) of the image. Large arrows indicate the directions in which lattice fringes give rise to displacements by phason strains. Small arrows show short-range periodical arrangements of image contrast.

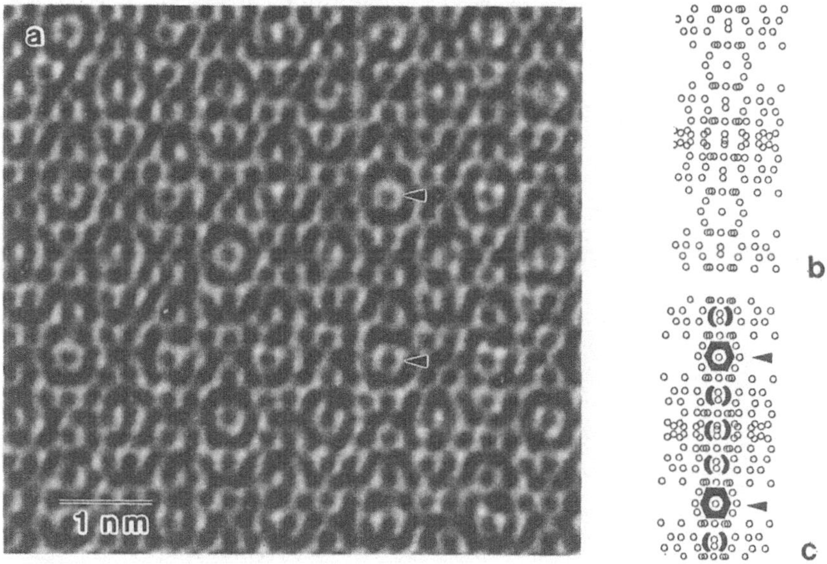

Fig. 10 (a): enlarged picture of a part of Fig. 9 (a). (b): atomic arrangement in the rectangle of Fig. 6(b); i.e. the projected atomic arrangement of Fig. 4(b) is placed at the twelvefold sites. (c): image contrast expected from the atomic arrangement of (b). Black regions in (c) correspond to bright regions in the observed image (a).

Acknowledgments

I sincerely thank Dr. A. Yamamoto, National Institute for Research in Inorganic Materials, for offering the diagrams of Figs. 6(a) and 6(b). This work has been supported by a Grant-in-Aid for Scientific Research on Priority Areas, "Quasicrystals", from the Ministry of Education, Science and Culture in Japan.

References

1) For example: K. Hiraga, M. Hirabayashi, A. Inoue and T. Masumoto: J. Microscopy 146, 245 (1987).
2) K. Hiraga and M. Hirabayashi: J. Electron Microsc. 36, 353 (1987).
3) K. Hiraga and M. Hirabayashi: Jpn. J. Appl. Phys. 26, L155 (1987).
4) K. Hiraga: Mat. Res. Soc. Symp. Proc. 139, 125 (1989).
5) C. L. Henley and V. Elser: Phil. Mag. B 53, L59 (1986).
6) P. Guyot and M. Audier: Mater. Sci. Forum 22-24, 247 (1987).
7) C. L. Henley: Phys. Rev. B 34, 797 (1986).
8) A. Yamamoto and K. Hiraga: Phys. Rev. B 37, 6207 (1988).
9) S. Nanao, Y. Tanaka, Y. Sakurai, C. Kokubu, Y. Watanabe, M. Masuda and I. Yasui: Mater. Sci. Eng. 99, 427 (1988).

Part IV

Quasiperiodic Lattices and Random Tiling

Icosahedral and Decagonal Quasicrystals in Al-Ni-M and Al-Cu-M (M = Transition Metal) Systems

A. Inoue[1], *An-Pang Tsai*[2], *and T. Masumoto*[1]

[1]Institute for Materials Reserach, Tohoku University, Sendai 980, Japan
[2]Graduate School, Tohoku University, Sendai 980, Japan

Abstract. Rapidly solidified phases in Al-Cu-M and Al-Ni-M ternary alloys were found to change, with the group number of M metal, in the sequence of an amorphous phase, followed by an icosahedral (I) phase, a decagonal (D) phase and a crystalline phase. It was furthermore found that the I phase in $Al_{65}Cu_{20}(Fe,Ru\ or\ Os)_{15}$ alloys and the D phase in $Al_{65}Cu_{15}(Co\ or\ Rh)_{20}$ and $Al_{70}Ni_{15}(Co\ or\ Rh)_{15}$ alloys were thermodynamically stable and existed in narrow composition ranges where the criteria of the outer electron per atom ratio $(e/a) \simeq 1.75$, the atomic size factor $(\lambda) \simeq 0.10$ and the ratio of the electronic structure parameters $K_p/2k_F \simeq 1.0$ are satisfied simultaneously. It is therefore interpreted that the formation and stability of the I and D phases in the Al-Cu-M and Al-Ni-M alloys are enhanced when the energy gap lies near the Fermi surface.

1. Introduction

Since the discovery of a metastable I phase in a rapidly solidified $Al_{86}Mn_{14}$ alloy[1], a number of metastable I alloys have been found in Al base systems containing various kinds of solute elements prepared by rapid solidification[2]. However, little has been known about the criteria for the formation of the I quasicrystal before stable I and D phases were discovered in a series of Al-Cu-M alloys[3]. Subsequently, similar stable D and metastable I phases were also found to be formed in an analogous Al-Ni-M system[4,5]. Accordingly, the Al-Cu-M and Al-Ni-M alloys seem to give a unique opportunity to derive the criteria for the formation and stability of the I and D quasicrystals. This paper is intended to present the changes of I and D phases with M elements in rapidly solidified and annealed Al-Cu-M and Al-Ni-M alloys and to investigate the reason why the stability of the I and D phases is enhanced at limited alloy compositions.

2. Experimental Details

The specimens used in the present work were Al-Cu-M and Al-Ni-M (M= transition metal) ternary alloys ranging from 15 to 20 %Cu or Ni and 10 to 20 %M. The subscripts represent the nominal atomic percentages. Mixtures of pure metals were melted in an argon atmosphere using an arc furnace. Ribbon samples of about 0.02 mm thickness and 1 mm width were prepared from these mixed alloys by a single roller melt-spinning apparatus. The amorphicity and quasicrystallinity of the samples in as-quenched, conven-

Springer Series in Solid-State Sciences, Vol. 93 Quasicrystals
Editors: T. Fujiwara · T. Ogawa

tionally solidified and annealed states were examined by various techniques of X-ray diffractometry, differential scanning calorimetry (DSC), differential thermal analysis (DTA), transmission and scanning electron microscopy (TEM, SEM) and optical microscopy (OM).

3. Results

The X-ray powder-diffraction patterns are shown in Fig. 1 for rapidly solidified $Al_{75}Cu_{15}V_{10}$, $Al_{65}Cu_{20}Cr_{15}$ and $Al_{65}Cu_{20}Mn_{15}$ alloys and in Fig. 2 for rapidly solidified $Al_{65}Cu_{20}Fe_{15}$ and $Al_{65}Cu_{15}Co_{20}$ alloys. As indexed in Figs. 1 and 2, the diffraction patterns reveal the formations of an amorphous phase in the Al-Cu-V alloy, a mostly single I phase in the Al-Cu-Cr, Al-Cu-Mn and Al-Cu-Fe alloys, and a mostly single D phase in the Al-Cu-Co alloy. Here, identification of the X-ray diffraction peaks was made by using six independent Miller indices as proposed by Bancel et al.[6] for the I phase and by Takeuchi and Kimura[7] for the D phase. Similar I and D phases were also found in rapidly solidified $Al_{65}Cu_{20}$(Ru or Os)$_{15}$ and $Al_{65}Cu_{15}Rh_{20}$ alloys, respectively. Considering the alloy compositions at which the mostly single I and D phases are formed, the I and D phases appear to have stoichiometric compositions which are approximated by $Al_{13}Cu_{4}M_{3}$ and $Al_{13}Cu_{3}M_{4}$, respectively. In addition, an amorphous phase was also formed in the rapidly solidified Al-Cu-M (M=Y, La, Ti, Zr or Hf) alloys. These results indicate that the as-quenched structure changes in the sequence of amorphous, I and D phases with an increase of the periodic group number of the M elements and is independent of their periodicity.

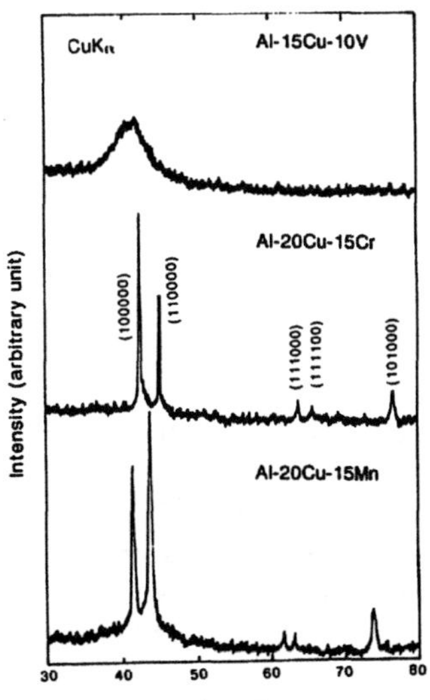

Fig. 1 X-ray powder-diffraction patterns of rapidly solidified $Al_{75}Cu_{15}V_{10}$ and $Al_{65}Cu_{20}$(Cr or Mn)$_{15}$ alloys.

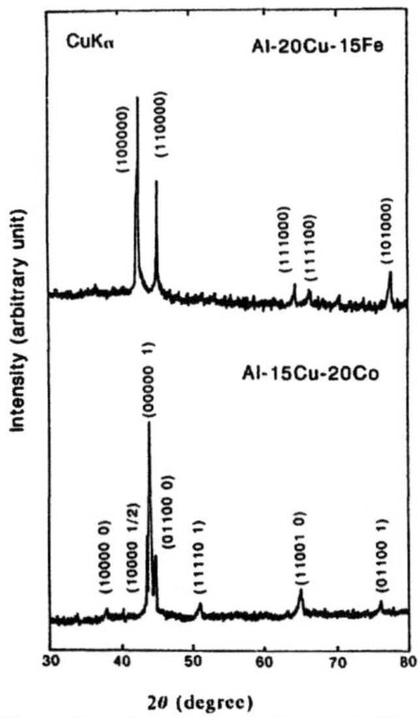

Fig. 2 X-ray powder-diffraction patterns of rapidly solidified $Al_{65}Cu_{20}Fe_{15}$ and $Al_{65}Cu_{15}Co_{20}$ alloys.

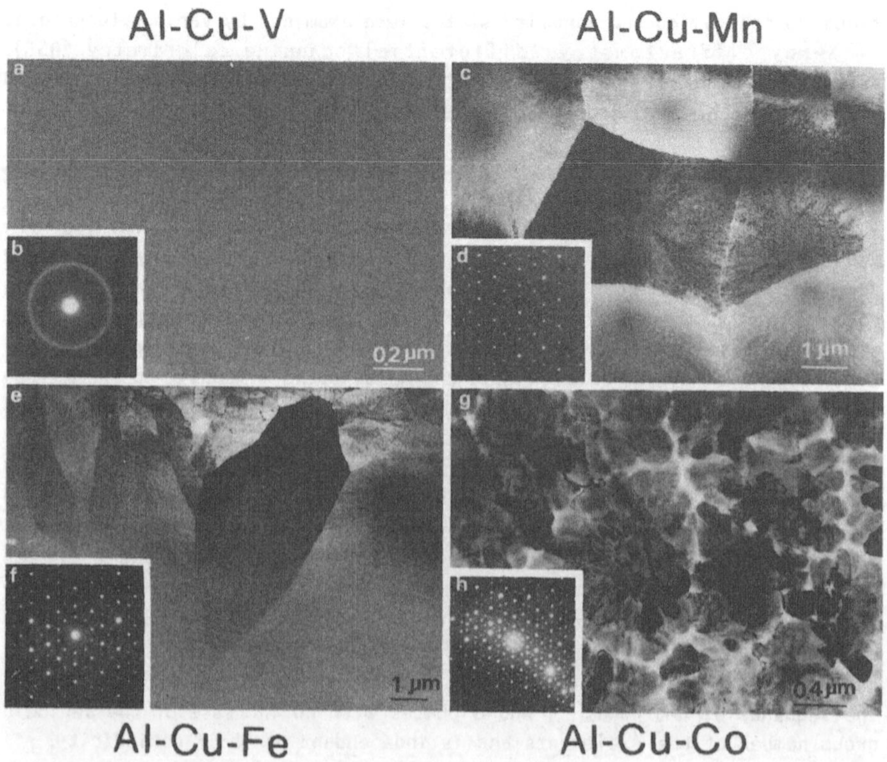

Fig. 3 Bright-field electron micrographs and selected area diffraction patterns of rapidly solidified $Al_{75}Cu_{15}V_{10}$, $Al_{65}Cu_{20}$(Mn or Fe)$_{15}$ and $Al_{65}Cu_{15}Co_{20}$ alloys.

Figure 3 shows bright-field electron micrographs and selected area diffraction patterns of rapidly solidified $Al_{75}Cu_{15}V_{10}$ (a and b), $Al_{65}Cu_{20}Mn_{15}$ (c and d), $Al_{65}Cu_{20}Fe_{15}$ (e and f) and $Al_{65}Cu_{15}Co_{20}$ (g and h). No contrast corresponding to a crystalline phase is observed in the Al-Cu-V alloy and the diffraction pattern consists only of several halo rings, revealing the formation of a mostly single amorphous phase. On the other hand, the diffraction patterns taken from the Al-Cu-Mn and Al-Cu-Fe alloys indicate a five-fold symmetry and are indexed as $[100000]_i$. The bright-field micrographs reveal that the I phase consists of large grains with an average size of about 20 µm for $Al_{65}Cu_{20}Mn_{15}$ and about 10 µm for $Al_{65}Cu_{20}Fe_{15}$. Additionally, the contrast revealing radiating branches is seen in their I alloys, indicating that the nucleation and growth of the I structure stemmed from a site within the grains and relatively large magnitude of phason strains generated in the formation of the I phase. Neither equiaxed subgrain boundaries nor clearly appreciable second phase is seen. It is to be noted that the grain sizes of the $Al_{65}Cu_{20}M_{15}$ (M=Mn or Fe) quasicrystals are 20 to 100 times as large as those (0.2 to 0.5 µm) of Al-Mn[8,9] and Al-Cr[10] quasicrystals prepared in the same quenching condition. Accordingly, growth of the I grains from liquid is thought to

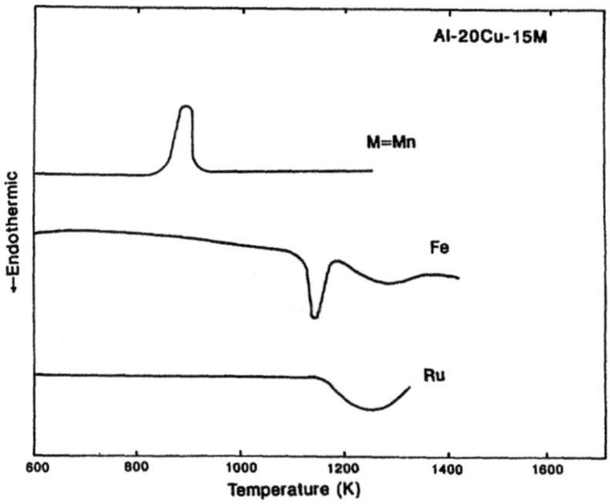

Fig. 4 Differential thermal analysis curves of I $Al_{65}Cu_{20}(Mn, Fe\ or\ Ru)_{15}$ alloys prepared by rapid solidification.

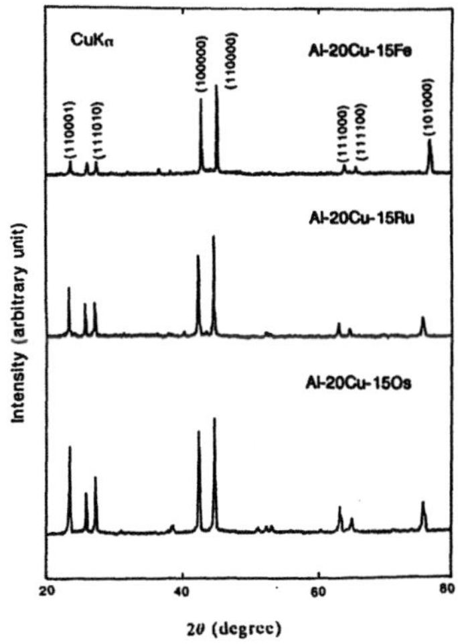

Fig. 5 X-ray powder-diffraction patterns of I $Al_{65}Cu_{20}(Fe, Ru\ or\ Os)_{15}$ alloys annealed for 12 h at 1130 K after conventional solidification.

be much faster for the Al-Cu-M alloys. Considering the large grain sizes and no trace of second phase for the I $Al_{65}Cu_{20}M_{15}$ (M=Mn or Fe) alloys, the formation of the I $Al_{13}Cu_4M_3$ phase is concluded to be much easier than that for almost all the Al-based quasicrystals reported to date. Additionally, Fig. 3 (g and h) shows the formation of the D $Al_{65}Cu_{15}Co_{20}$ alloy in a rapidly solidified state. The diffraction pattern reveals the tenfold symmetry characteristic of the D structure. The D phase consists of grains with a size of 1 to 2 μm and each grain contains a high density of subboundaries, being in contrast to the result[11] that the conventionally solidified and fully annealed $Al_{65}Cu_{15}Co_{20}$ alloys have a D phase without any trace of subboundaries and the grain size is as large as over 2 μm.

The transformation behavior of the quasicrystals to equilibrium phases was examined by DSC and DTA. Figure 4 shows the DTA curves of the I $Al_{65}Cu_{20}M_{15}$ (M=Mn, Fe or Ru) alloys measured at a heating rate of 10 K/min. Only the I $Al_{65}Cu_{20}Mn_{15}$ alloy exhibits a large exothermic peak resulting from the transformation of the I phase to a crystalline phase in a temperature range of 783 to 858 K. On the other hand, the I Al-Cu-Fe and Al-Cu-Ru alloys do not show any exothermic peak at temperatures below 1200 K, indicating that the I phase in Al-Cu-Fe and Al-Cu-Ru systems has high thermal stability. The high thermal stability allows expectation of obtaining thermodynamically stable I Al-Cu-Fe and Al-Cu-Ru alloys.

Figure 5 shows the X-ray powder-diffraction patterns of $Al_{65}Cu_{20}Fe_{15}$, $Al_{65}Cu_{20}Ru_{15}$ and $Al_{65}Cu_{20}Os_{15}$ alloys annealed for 12 h at a high temperature of 1130 K after conventional solidification. Most of the diffraction peaks can be identified as the I phase, as indexed in Fig. 5. This clearly indicates that the I phase is formed as a thermodynamically

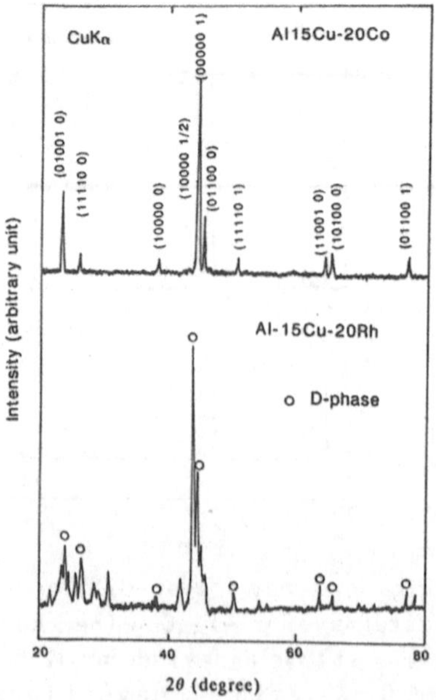

Fig. 6 X-ray powder-diffraction patterns of D $Al_{65}Cu_{15}$(Co or Rh)$_{20}$ alloys annealed for 36 h at 1200 K after conventional solidification.

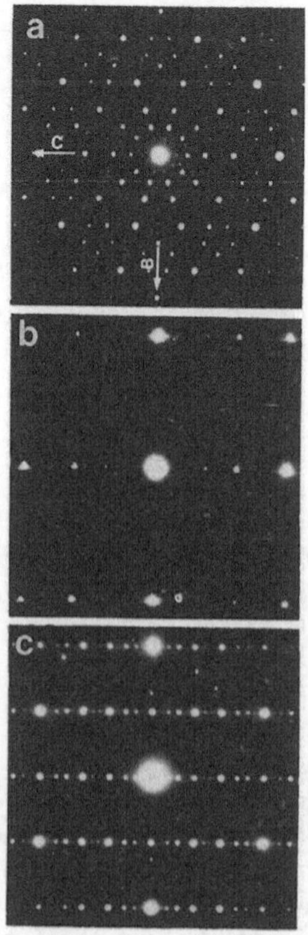

Fig. 7 Electron diffraction patterns of a D $Al_{65}Cu_{15}Co_{20}$ alloy annealed for 36 h at 1200 K after conventional solidification.

stable phase in the three alloys. The lattice parameters of the Al-Cu-Ru and Al-Cu-Os quasicrystals determined from the X-ray diffraction data are larger by about 1.5 % than that of the Al-Cu-Fe quasicrystal because the atomic sizes of Ru and Os are 6.0 to 6.7 % larger than that of Fe.

In addition, Fig. 6 shows the X-ray powder-diffraction patterns of the $Al_{65}Cu_{15}Co_{20}$ and $Al_{65}Cu_{15}Rh_{20}$ alloys annealed for 36 h at 1200 K after conventional solidification. All diffraction peaks in the Al-Cu-Co alloy and main diffraction peaks in the Al-Cu-Rh alloy can be indexed as the D phase, revealing that the D phase is also formed as a stable phase in their alloys. In order to clarify more clearly the two-dimensional quasicrystalline structure in the stable D Al-Cu-Co alloy, electron diffraction patterns which were taken at different incidences are shown in Fig. 7. The diffraction pattern (Fig. 7(a)) was taken at an incidence along the tenfold axis and the patterns (Fig. 7(b) and (c)) correspond to those taken at the incidences perpendicular to B and C in Fig. 7(a). The

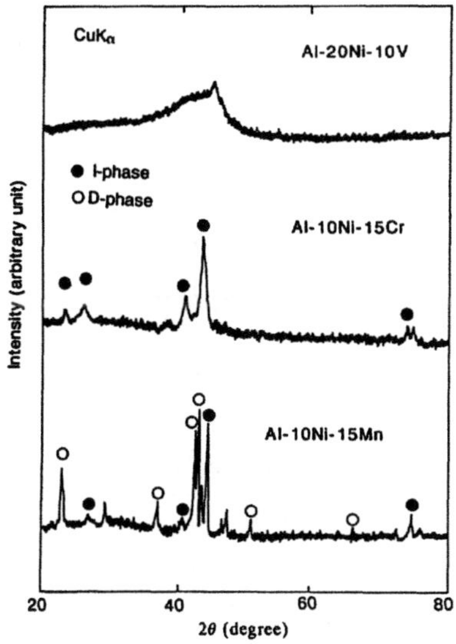

Fig. 8 X-ray powder-diffraction patterns of rapidly solidified Al-Ni-M (M=V, Cr or Mn) alloys.

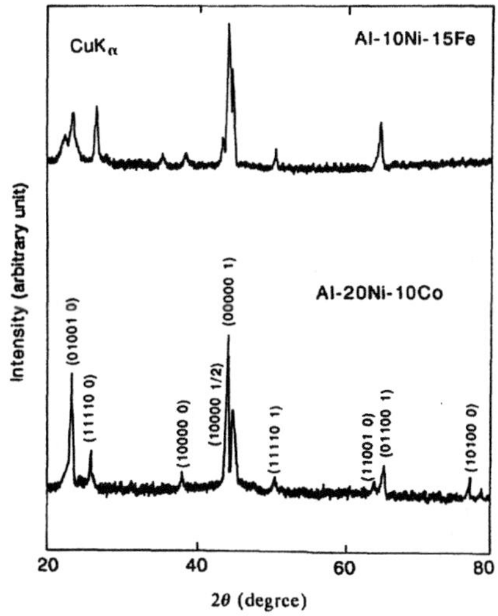

Fig. 9 X-ray powder-diffraction patterns of rapidly solidified Al-Ni-M (M=Fe or Co) alloys.

reflection spots along the vertical and horizontal directions in Fig. 7(b) and (c) are characterized by a periodic array and by an arrangement at the golden ratio, respectively, indicating the existence of periodicity along one direction. On the other hand, Fig. 7(a) reveals a tenfold symmetry for the D structure taken perpendicular to the periodic axis. The results obtained from X-ray diffractometry and TEM observation allow us to confirm the existence of the stable D phase in the Al-Cu-Co alloy. Here, it appears important to describe that the formation of the stable D phase is limited to the narrow range around $Al_{65}Cu_{15}Co_{20}$, though the quenching-induced D phase is formed in a wider composition range of 12 to 22 %Cu and 13 to 23 %Co[12].

It was described that the as-quenched phase in the Al-Cu-M (M=Y, La, Ti, Zr, Hf or V) alloys was composed of an amorphous phase. The amorphous phase only in the $Al_{75}Cu_{15}V_{10}$ alloy was found[13,14] to transform into an I single phase accompanied by an exothermic heat of 1.37 kJ/mol in the temperature range of 704 to 750 K, when the sample is heated at a rate of 40 K/min. The resulting I phase consists of grains with a size of 0.3 to 0.5 μm containing a high density of internal defects, but no second phase is seen on the grain boundaries. It is therefore concluded that the Al-Cu-V alloy has two different metastable structures of amorphous and I phases which are dependent on the preparative condition.

A similar sequent change in rapidly solidified structure with the group number of M metal was observed in Al-Ni-M ternary alloys. The X-ray diffraction pattern of a rapidly solidified phase is shown in Fig. 8 for $Al_{70}Ni_{20}V_{10}$, $Al_{75}Ni_{10}Cr_{15}$ and $Al_{75}Ni_{10}Mn_{15}$ and in Fig. 9 for $Al_{75}Ni_{10}Fe_{15}$ and $Al_{70}Ni_{20}Co_{10}$. As identified in these figures, the structure consists

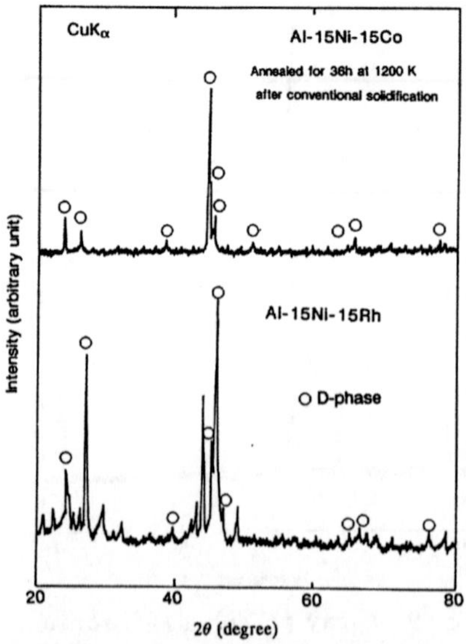

Fig. 10 X-ray powder-diffraction patterns of D $Al_{70}Ni_{15}(Co\ or\ Rh)_{15}$ alloys annealed for 36 h at 1200 K after conventional solidification.

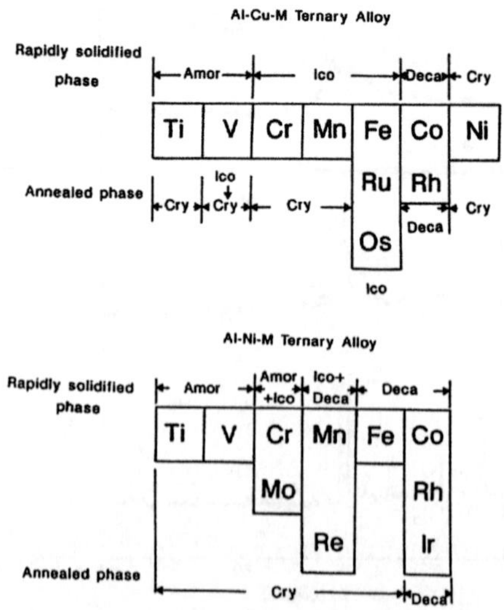

Fig. 11 Change in the rapidly solidified and annealed phases in Al-Cu-M and Al-Ni-M (M=Ti, V, Cr, Mo, Mn, Re, Fe, Ru, Os, Co or Rh) ternary alloys with M metal.

of a mostly amorphous phase for the Al-Ni-V alloy, coexistent amorphous and I phases for the Al-Ni-Cr alloy, coexistent I and D phases for the Al-Ni-Mn alloy and a mostly single D phase for the Al-Ni-Fe and Al-Ni-Co alloys. Although the D phase in the Al-Ni-Fe system decomposes to a crystalline phase at about 930 K, the D Al-Ni-Co phase prepared by rapid solidification remained stable up to the melting temperature. A thermodynamically stable D phase was searched in a fully annealed state after conventional solidification and both $Al_{70}Ni_{15}Co_{15}$ and $Al_{70}Ni_{15}Rh_{15}$ alloys are composed of a stable D phase as shown in Fig. 10. The stable D phase has been clarified[11] to have a unique growth morphology consisting of columns with a cross section of D shape which is just the same as that[5,11] for the stable Al-Cu-(Co or Rh) D alloys. These above-described data clearly indicate that the structural change in Al-Ni-M ternary alloys with M metal takes place in the sequence of amorphous, amorphous plus metastable I, metastable I plus metastable D, metastable D, and stable D phases.

On the basis of the above-described results, the changes of as-quenched and annealed phases in Al-Cu-M and Al-Ni-M alloys with M metal are summarized in Fig. 11. One can notice a systematic change in the as-quenched and annealed structures with the periodic group number of the M metals. That is, the as-quenched structure is the amorphous phase for both alloys containing the M elements of the IIIA, IVA and VA groups, the metastable I phase for the Al-Cu-M alloys containing the VIA and VIIA elements, the metastable amorphous plus metastable I, and metastable I plus D phases for the Al-Ni-M alloys containing the VA and VIA elements, the stable I phase for the Al-Cu-M alloys containing Fe, Ru and Os in the VIIIA group and the stable D phase for the Al-Cu-M and Al-Ni-M alloys containing Co and Rh in the same VIIIA group. Furthermore, the amorphous Al-Cu base alloy containing V which is located at the boundary between amorphous and metastable I phases changes into the metastable I phase by annealing for 5 min at 820 K, followed by the transition to crystalline phases after annealing for 48 h at 820 K. In addition, the Al-Cu-M alloys containing Ni, Pd or Pt element in the VIIIA group are composed of crystalline phases even in an as-quenched state. Accordingly, as shown in Fig. 8, the structural change with an increase of the periodic group number of the M elements occurs in the order of amorphous, followed by amorphous plus metastable I, metastable I, stable I, stable D and then crystalline phases for the Al-Cu-M alloys and in the order of amorphous, followed by amorphous plus metastable I, metastable I plus metastable D, metastable D and stable D phases for the Al-Ni-M alloys.

4. Discussion

The reason for the high stability of the I structure in Al-Cu-M (M=Fe, Ru or Os) and Al-Ni-M (M=Fe or Ru) as well as the D structure in Al-Cu-M and Al-Ni-M (M=Co or Rh) will be discussed in this section. In general, the structural stability of alloys is mainly dominated by electronegativity, atomic size and energy-band structure. The electronegativity and the atomic size have a significant influence on the short-range atomic configuration. However, considering the crystal structure on a long range scale, it is necessary to take the role of the energy-band structure into consideration. Accordingly, consideration on the basis of the energy band is important for metallic materials where conductive electrons do not localize. Here, the reasons for the formation of the I and D phases in Al-Cu-M (M=V, Cr, Mn, Fe, Ru, Os, Co or Rh) and Al-Ni-M (M=Mn, Fe, Ru, Co or Rh) systems and for the high stability of the I Al-Cu-(Fe, Ru or Os) and D Al-Cu-(Co or Rh) and Al-Ni-(Co or Rh) alloys are investigated in terms of the three parameters which dominate the structural stability of alloys. In alloy systems with large differences in electronegativity, the electronegativity seems to play an important role in the structural stability of alloys. On the contrary, the factor can be ignored for the alloys where the constituent elements have small differences in electronegativity. Since the difference in the electronegativity is small in the Al-Cu-M and Al-Ni-M alloys, its contribution is ignored, in agreement with the treatment for the I Al-Mn alloys by Bancel and Heiney[15]. Accordingly, the formation and high structural stability of the I and D phases in Al-Cu-M and Al-Ni-M systems are investigated in terms of the two factors of energy-band structure and atomic size.

Hume-Rothery[16] has pointed out an empirical rule in which the crystal structure in some noble metal-base binary alloys is dominated by the concentration of outer electrons. Here, the outer electron concentration presents the number of conductive electrons per one atom and is usually expressed by e/a. Subsequently, Jones[17] tried to explain the empirical Hume-Rothery rule from the interaction between the Fermi sphere containing the valence electrons and the Brillouin zone boundaries. On the basis of the concept proposed by Pauling[18] in which the transfer of electrons takes place from the conduction band to the d-band so as to fill an unpaired electronic state in transition metals, the e/a values of each transition metal are estimated[19] to be -4.66 for Cr, -3.66 for Mn, -2.66 for Fe, -1.71 for Co and -0.61 for Ni. In addition, the atomic size factor (λ) was evaluated[20] from the solute concentrations (C_b and C_c) of each nonequilibrium phase by eqs. (1), (2) and (3).

$$\Delta V_{ab}/V_a=(V_b-V_a)/V_a=(r_b/r_a)^3-1, \quad (1)$$
$$\Delta V_{ac}/V_a=(V_c-V_a)/V_a=(r_c/r_a)^3-1, \quad (2)$$

such that

$$\lambda=|\Delta V_{ab}/V_a|.C_b + |\Delta V_{ac}/V_a|.C_c \quad (3)$$

Here, V_a and r_a are the atomic volume and atomic size of the matrix and V_b and r_b are the atomic volume and atomic size of the solute elements.

Figure 12 shows the relationship between λ and e/a for the I and D Al-Cu-M and Al-Ni-M alloys, along with the data on Al-M alloys belonging to the Mackay-type I structure and Al-Li-Cu, Mg-Cu-Zn, Mg-Al-M (M=Cu, Zn or Ag) alloys belonging to the Frank Kasper (FK)-type I structure. The FK-type I alloys form in the e/a range of 2.0 to 2.2, in agreement with the proposal by Villars et al.[21], while the Mackay-type I Al-Cu-M and Al-Ni-M alloys form in the e/a range of 1.6 to 1.8. Thus, the difference in e/a between the Al-(Cu or Ni)-M alloys and the FK-type alloys is thought to result from the difference in their structures. The mostly single I alloys in Al-M, Al-Cu-M and Al-Ni-M systems form in the ranges of 1.6 to 1.8 for e/a and 0.06 to 0.11 for . The λ values are smaller for the Al-Cr and Al-Mn alloys and the replacement of Al by Cu, Ni, Si or Ge gives an increase of λ, leading to the increase of the grain size and the quasicrystallinity. Accordingly, it is said that the formation of the I and D phases with better quasicrystallinity, higher formation tendency and higher stability is possible in the simultaneous satisfaction of the two criteria of about 1.75 for e/a and 0.08 to 0.10 for λ.

Nagel and Tauc[22] proposed in the frame of a nearly free-electron model that the electronic energy decreases, leading to the stabilization of the amorphous phase, when the criterion of $K_p=2k_F$ is satisfied, i.e., when the Fermi level is at a minimum in the density of electronic states. Here, K_p is the diameter of the Brillouin zone boundary and $2k_F$ is the diameter of the Fermi sphere. Their proposal suggests that the amorphous phase has the highest stability at the composition where the static structure factor S(k) has its first peak at $K_p=2k_F$. There are some experimental data[23,24] that the density of electronic states has a minimum value when the electronic energy is nearly equal to the Fermi energy E_F, resulting in the high stability of amorphous phase. Accordingly, it seems reasonable to apply this concept to the I and D Al-Cu-M and Al-Ni-M alloys. The intense reflection peaks, $(100000)_i$, $(110000)_i$ and $(10000\ 1)_d$, in the X-ray diffraction patterns of the I and D quasicrystals correspond to the most intense halo peak for an amorphous phase and enable us to evaluate the K_p value. Furthermore, the $2k_F$ value can be calculated from atomic density

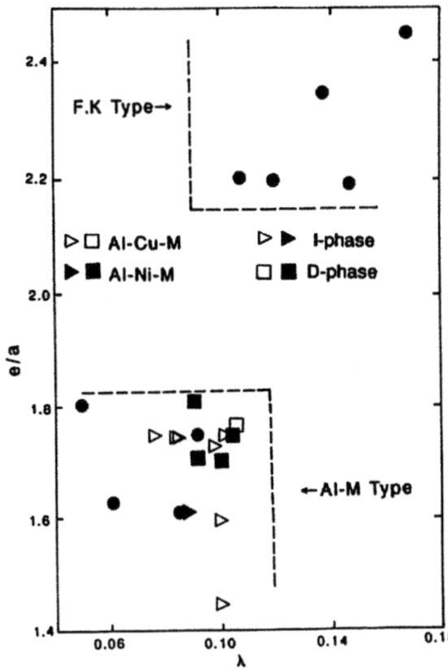

Fig. 12 Relationship between λ and e/a for the I and D phases in Al-Cu-M and Al-Ni-M (M=V, Cr, Mn, Re, Fe, Ru, Os, Co or Rh) alloys. The data on the other quasicrystals are also shown for reference.

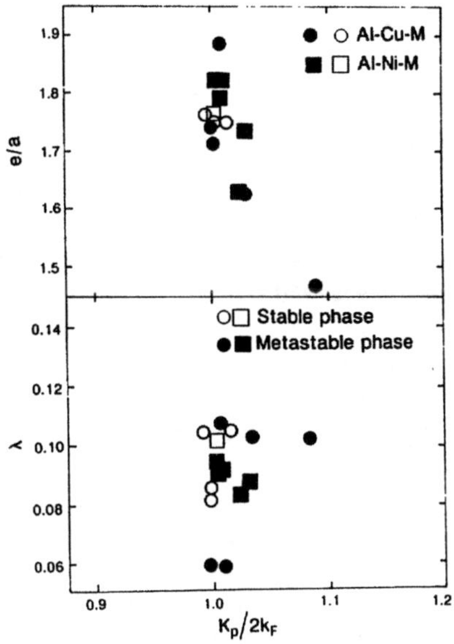

Fig. 13 Relationship between $K_p/2k_F$ and e/a or λ for the I and D phases in Al-Cu-M and Al-Ni-M (M=V, Cr, Mn, Re, Fe, Ru, Os, Co or Rh) alloys.

and e/a in the frame of the free electron model. Figure 13 shows the relationship between $K_p/2k_F$ and e/a or λ for the Al-(Cu or Ni)-M quasicrystalline alloys. It is seen that the stable I and D alloys lie in the vicinity of $K_p/2k_F \simeq 1.0$, even though the metastable I and D alloys have relatively scattered values of $K_p/2k_F$. This indicates that the formation tendency and stability of the Al-Cu-M and Al-Ni-M quasicrystals are enhanced when the energy gap is located in the vicinity of the Fermi surface. That is, it is presumed that the $Al_{65}Cu_{20}$(Fe, Ru or Os)$_{15}$, $Al_{65}Cu_{15}$(Co or Rh)$_{20}$ and $Al_{70}Ni_{15}$(Co or Rh)$_{15}$ compositions are in the lowest state of electronic energy probably because of the structure which places the Brillouin zone boundaries in contact with the Fermi surface. It is finally important to point out that the presumption derived from the relationship shown in Fig. 13 is consistent with the recent theoretical expectation[25] in which standard elementary models derived from the Hume-Rothery rules are valid in the explanation of the formation of metallic quasicrystals at limited alloy compositions. However, the reason why the distinction between the stable I and the stable D phases takes place in the same criteria of $K_p/2k_F \simeq 1.0$, $e/a \simeq 1.75$ and $\lambda \simeq 0.10$ is under investigation.

References

1. D. Shechtman, I. Blech, D. Gratias and J.W. Cahn: Phys. Rev. Lett., 54(1984), 1951.
2. C. Suryanarayana and H. Jones: Int. J. Rapid Solidif., 3(1988), 253.
3. A.P. Tsai, A. Inoue and T. Masumoto: Mater. Trans., Japan Inst. Met., 30(1989), 666.
4. A.P. Tsai, A. Inoue and T. Masumoto: Mater. Trans., Japan Inst. Met., 30(1989), 150.
5. A.P. Tsai, A. Inoue and T. Masumoto: Mater. Trans., Japan Inst. Met., 30(1989), 463.
6. P.A. Bancel, P.A. Heiney, P.W. Stephens, A.I. Goldmans and P.M. Horn: Phys. Rev. Lett., 54(1985), 2422.
7. S. Takeuchi and K. Kimura: J. Phys. Soc. Japan, 56(1987), 982.
8. D. Shechtman and I.A. Blech: Met. Trans., 16A(1985), 1005.
9. T. Masumoto, A. Inoue, M. Oguchi, K. Fukamichi, K. Hiraga and M. Hirabayashi: Trans. Japan Inst. Metals, 27(1986), 81.
10. A. Inoue, H.M. Kimura and T. Masumoto: J. Mater. Sci., 22(1987), 1758.
11. A.P. Tsai, A. Inoue and T. Masumoto: Mater. Trans., Japan Inst. Metals, 30(1989), 300.
12. L.X. He, Y.K. Wu and K.H. Kuo: J. Mater. Sci., 7(1988), 1284.
13. A.P. Tsai, A. Inoue and T. Masumoto: Jpn. J. Appl. Phys., 26(1987), L1994.
14. A.P. Tsai, A. Inoue, Y. Bizen and T. Masumoto: Acta Met., 37(1989), 1443.
15. P.A. Bancel and P.A. Heiney: Phys. Rev. B, 33(1986), 7319.
16. W. Hume-Rothery: J. Inst. Metals, 35(1926), 295.
17. H. Jones: Pro. Phys. Soc. London, 49(1937), 250.
18. L. Pauling: Phys. Rev., 54(1938), 899.
19. G.V. Raynor: Progress in Metal Phys., 1(1949), 1.
20. S. Ueno and Y. Waseda: Sci. Rep. Res. Inst. Tohoku Univ., 32A(1984), 97.
21. P. Villars, J.C. Phillips and H.S. Chen: Phys. Rev. Lett., 57(1986), 3058.
22. S.R. Nagel and J. Tauc: Phys. Rev. Lett., 35(1975), 380.
23. D. Nicholson and L. Schwartz: Phys. Rev. Lett., 49(1982), 1050.
24. H.F. Banmann, J. Krieg, G. Indlekofer, P. Oelhafen and H.T. Guntherodt: Phys. Rev. Lett., 51(1983), 714.
25. J. Friedel: Helvetica Physica Acta, 61(1988), 538.

Growing Perfect Quasicrystals

*P.J. Steinhardt**

Institute for Advanced Study, Princeton, NJ 08540, USA
*On leave from the Department of Physics, University of Pennsylvania, Philadelphia, PA 19104, USA

Abstract: We review recently developed algorithms for aggregating tiles to form defect-free quasicrystal tilings. The algorithms show that, contrary to conventional wisdom, long range atomic interactions are not required to grow perfect quasicrystals; in particular, there exists a class of topological defects called "decapods" which can act as ideal seeds for rapid, perfect quasicrystal growth. We also discuss how "growability" via local interactions depends upon the quasiperiodic arrangement of unit cells (the *local isomorphism (LI) class*); consequently, only a small subclass of LI classes may be physically realizable, perhaps settling a serious ambiguity in the structural determination of atomic structure.

1 Introduction

Since the 1984 discovery of the first icosahedral alloy (Al_6Mn) by Shechtman *et al.* [1], more than one hundred icosahedral alloys have been found, including samples which exhibit substantially improved long-range atomic positional order. During the same period, there has been considerable controversy concerning the correct model to describe the structure of icosahedral alloys. Three basic models have been proposed: (1) multiple-twinning [2], (2) icosahedral glass [3], and (3) quasicrystal [4].

Two years ago, the status of the controversy was somewhat murky. Multiple-twinning models seemed implausible because unit cells containing tens of thousands, and, more recently, hundreds of thousands of atoms are required to obtain models in agreement with diffraction data from the most perfect samples. The icosahedral glass and quasicrystal models are more difficult to distinguish because both predict long-range icosahedral orientational order. The difference is that the icosahedral glass model predicts *no* long-range positional order whereas the quasicrystal model predicts long-range *quasiperiodic* positional order. The simplest icosahedral glass model is constructed by randomly packing icosahedral units face-to-face or vertex-to-vertex. The randomness results in only short-range positional order so that diffraction peaks should exhibit an irreducible finite width associated with the disorder caused by the random packing. The quasicrystal model, by contrast, predicts true Bragg peaks for a perfect sample. The experimental observation of finite peak widths can be explained as due to defects and strains quenched into the sample during solidification. A year and a half ago, the high degree of positional order observed in the best samples of $AlLiCu$ ruled out the simplest icosahedral glass models [5]. Hence, the quasicrystal model (with allowance for defects frozen-in during solidification) appeared to emerge as the simplest model for describing icosahedral alloys [6].

Nevertheless, there remained a great reluctance to accept the quasicrystal picture due to a fundamental conceptual difficulty — a fatal flaw according to some. It seemed apparent that long-range atomic interactions are required to grow perfect quasicrystals. Consider an atom impinging on the surface of a *quasicrystal* seed cluster. The sequence of atoms in the seed cluster never exactly repeats, so it appears that the atom must interact

Springer Series in Solid-State Sciences, Vol. 93 **Quasicrystals**
Editors: T. Fujiwara · T. Ogawa © Springer-Verlag Berlin, Heidelberg 1990

with each atom in the seed cluster to ensure that it sticks at a site consistent with perfect *quasiperiodic* order. As the seed grows, this condition requires arbitrarily long-range interactions, which is physically implausible. With a finite-range interaction, a high density of defects seems inevitable. For several years, the theoretical argument resonated strongly with the observations of experimentalists. A generic complaint was that all icosahedral alloys exhibit residual disorder as indicated by finite diffraction peak widths. In light of the theoretical argument and experimental experience, it seemed possible that the ideal quasicrystal is a physically unreachable state.

As reported at this symposium, two important changes have taken place. First, it has been conclusively shown that that the theoretical assessment is incorrect: there is no barrier to growing perfect quasicrystals using only finite-range atomic interactions [7]. Second, an experimental breakthrough has taken place in which significantly more perfect icosahedral alloys have been discovered. The new alloys, $AlFeCu$ and $AlRuCu$, exhibit near-perfect diffraction patterns [8]. Together, these changes remove the remaining objections to the quasicrystal picture.

The purpose of this article is to briefly recount the recent theoretical progress in the theory of quasicrystal growth, work obtained in collaboration with G. Onoda, D. DiVincenzo, J. Socolar and K. Ingersent. In this discussion, I will suppress most mathematical details, especially the proofs of the results, which can be found in the principal references [7] and [9]. My intent is for this paper to serve as a useful prologue that highlights some of the subtleties and surprises which might otherwise be lost in the highly technical discussions.

2 More Arguments for Long-Range Interactions

The claim that long-range atomic interactions are required to grow perfect quasicrystals is supported by more than just the heuristic argument given in the previous section. First, it can be rigorously proven that long-range interactions are required to grow a quasiperiodic chain in 1D. For example, consider "growing" via a *local rule* a 1D chain consisting of two types of atoms, A and B, arranged in a Fibonacci sequence [4], $\ldots ABABBAB\ldots$. By a local rule, I mean that, in deciding whether to add A or B to the end of a finite chain, only the sequence of atoms within some finite distance of the end of the chain should be ascertained, as might be mimicked by some local atomic interaction. Unfortunately, there are certain points along the chain where the information required to make the correct choice for the next addition to the chain includes the entire sequence of atoms stretching to the other end of the chain. An example is the finite chain

$$BABABBABABBAB.$$

For the next two atoms on the right-hand side, there are two apparent possibilities, AB or BA (two A's or three B's never occur in succession occurring in a perfect chain). However, the first choice, AB is actually illegal upon careful inspection. For this choice the sequence would consist of $BABAB$ repeated exactly three times in succession, whereas ideal Fibonacci chains do not contain three repetitions of any subsequence in succession. However, to recognize the problem requires knowledge of the entire sequence, including the two elements on the left-hand end of the chain. In the context of our discussion, then, this implies that interactions extending over the length of the chain are required to ensure continued growth of a perfect Fibonacci chain. However, owing to the self-similarity of the chain, we could exhibit chains of arbitrary length which display the same property. Hence, it becomes obvious that infinite-range interactions are necessary. The result can be extended to quasiperiodic structures in 2D and 3D with allowed crystallographic symmetry (i.e., incommensurate crystals). Such structures have quasiperiodic order along two or three independent axes, respectively. (Incommensurate crystals exist in nature, nevertheless, but this is probably because they can be obtained by growing two independent crystals lattices which are rotated and/or stretched with respect to one another. Each crystal lattice can be grown by local interactions which extend over the length of the unit cell. There is no analogue for the quasicrystals.)

A reasonable conjecture would be that quasicrystals are, if anything, a worse problem. For example, the 2D Penrose tiling [10] can be decorated with Ammann lines which form *five* independent sets of parallel lines spaced according to a Fibonacci sequence of long and short intervals [4]. In "growing" a tiling from some seed cluster, the sequence of Ammann lines must be extended in each of five directions simultaneously. It seems that the 1D growth problem multiplies by five into a Penrose nightmare!

A potential mitigating factor is the Penrose edge-matching rule. Penrose developed a set of a edge-matching rules that constrain the way two tiles can match together edge-to-edge. (The rules are typically indicated by placing arrows on the edges of the tiles [11].) Penrose showed that the only plane-filling tiling consistent with the matching rules is a perfect Penrose tiling. Do these edge matching rules also represent viable local rules for "growing" a tiling by aggregation; e.g., adding one tile at a time to a random edge in accordance with the edge-matching rule? Absolutely not! Penrose's theorem guarantees that the Penrose tiling is the unique space-filling tiling consistent with the rules, but it does not represent a local algorithm for constructing the tiling. Instead, following this procedure, mistakes are made every ten tiles or so; each time, tiles have to be removed and a different choice made.

Based on the rigorous proof for 1D quasiperiodic chains and higher dimensional incommensurate crystals, and on the dismal failure of the Penrose edge-matching rules as growth rules for pentagonal quasicrystals, the common conclusion had been that long-range interactions are required to grow perfect quasicrystals.

3 How to Grow Perfect Penrose Tilings

Given the compelling arguments in the previous section, it is very surprising that the conclusion turns out to be *dead wrong!* In 1988, G. Onoda, D. DiVincenzo, J. E. S. Socolar and the author presented a tiling algorithm for growing perfect Penrose tilings which could be mimicked by a physical system with purely local atomic interactions [7]. In this section, we discuss the basic features, emphasizing the relation between the tiling algorithm and a realistic physical system.

The first issue is the range of interactions required to grow a perfect Penrose tiling (PPT). Recall that the Penrose edge-matching rules failed, and there was an heuristic argument suggesting that infinite-range rules are required. In fact, the requisite growth rules are remarkably short-range: edge-matching rules must be replaced by "vertex-matching rules." That is, in deciding whether to add a given tile to a given vertex on the surface of a seed cluster, it is only necessary to determine the configuration of tiles already adjoined to that site. The vertex-matching rule demands that the vertex configuration be consistent with one of the eight types that appear in a PPT [7]. The vertex-matching rules subsume the Penrose edge-marking rules; that is, any tiling obeying the vertex rules also obeys the edge rules. Surprisingly, the rule is only marginally longer-range than the Penrose edge-matching rule. By contrast, to grow a large unit cell, rational approximant tiling consisting of the same rhombus units as the Penrose tiling requires local rules extending to the size of the unit cell. This distance can be arbitrarily long for closer and closer rational

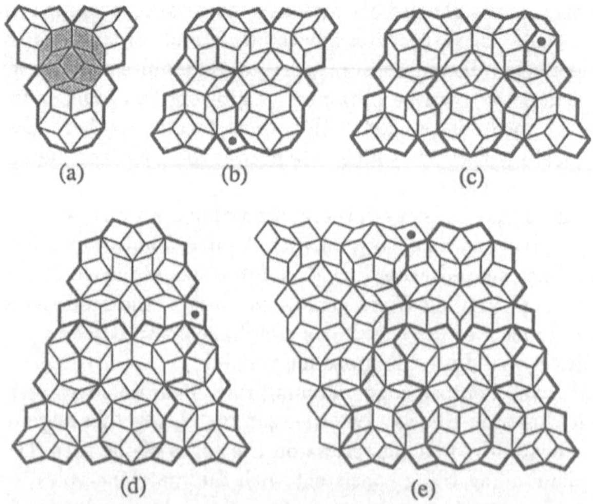

Figure 1: Growth beginning from a perfect seed cluster (shaded in (a)), highlighting dead surfaces encountered (heavy lines). A fat rhombus (black dot) is added according to Rule 2, resulting in new forced sites.

approximants, much longer range than the vertex-matching rules. The counter-intuitive conclusion is that it actually takes shorter range rules to grow a PPT than a large unit cell crystal approximant!

If we consider adding a tile to a free edge extending from an incomplete vertex, there may be only one choice (fat or skinny rhombus) consistent with the vertex-matching rules, in which case the site is said to be *forced.* Alternatively, either choice may be consistent, in which case the site is *unforced.* To grow a PPT beginning from a perfect seed cluster (a single tile or a cluster of tiles that occurs in a PPT):

1. Randomly choose any forced site on the surface and add a forced tile to it. Continue this process until there are no forced sites left. A surface with no forced sites is referred to as a *dead surface.* The dead surface consists of macroscopically straight edges joined at 108 or 72 degree corners. See Fig. 1.

2. Upon reaching a dead surface, randomly choose a corner and add a fat rhombus to it [12].

A reasonable first reaction is that these rules do not comprise a "local" algorithm [13]. Rule 2 is invoked only after a search of the complete surface determines that there are no forced sites. However, the effect of surface inspection can be implemented in a physically realistic stochastic growth process in which atoms interacting through local forces impinge randomly on surface sites and stick according to *locally* determined sticking probabilities. The rules governing forced, corner and all other sites can be mimicked by choosing sticking probabilities $p_f \gg p_c > 0$ and $p_o = 0$. The condition that $p_f \gg p_c$ causes growth to occur at forced sites before corner sites. The key point is that the sticking probability for a given site is determined only by the local arrangement of atoms around the vertex and can be set by appropriately chosen short-range interactions.

The notion of widely varying sticking probabilities and forced versus other sites is already familiar in periodic crystal growth. It is well known that the value of p along a free surface can vary widely depending on the presence of ledges, kinks, screw dislocations, corners and local atomic configurations. Since $p \propto \exp(-E_b/kT)$, where E_b is the binding energy, variations by factors as great as e^{100} are plausible.

The growth rules are recovered exactly in the limit $p_o = 0$ and as $p_f/p_c \to \infty$. The strict infinite limit is not realistic in the sense that it would require infinite time to grow the tiling; growth would be stymied for an arbitrarily long time at each dead surface in the limit that the sticking probability at a corner site becomes arbitrarily small. However, we can claim that *there can exist local interactions that ensure the growth of arbitrarily large PPTs with arbitrarily small probabilities of error.* For example, with the above rules, $p_f/p_c > N$ is sufficient to grow a PPT with order N tiles. Our most pessimistic estimate is that perfect growth from a perfect seed (see later discussion for growth from imperfect seeds) takes roughly $N^{(d-1)/d}$ times longer than to grow a perfect crystal. This is sufficiently rapid to grow macroscopic perfect grains (diameter equal to 10 μm, or 10^{13} atoms) on laboratory time scales.

We have begun to consider variants of the original rules that can change the morphology and speed of growth. The original rules in Reference 7 required that, upon reaching a dead surface, fat rhombus tiles only be added to 108 degree corners. The less restrictive rule described above (adding to any type of corner) results in no substantial difference. In both cases, a sequence of different dead surface shapes appears in the growth process. In general, four types of dead surfaces — rhombuses, trapezoids, coffin-shapes and double-coffin shapes [7] — appear in random order depending upon the previous dead surface and the random corner chosen for adding a fat tile. However, if we require that fat tiles only be added to 72 degree corners, a different morphology results. First, some seeds (less than 5%) cannot be grown at all because they reach dead surfaces that have no 72 degree corners. However, the remaining seeds can be grown ad infinitum. The new feature is that only one type of dead surface appears during growth. Each dead surface is a trapezoid of similar shape to the previous dead surface. For technical reasons, the constraints on sticking probabilities may be somewhat less for this growth rule.

An especially dramatic increase in growth speed can be achieved using a fascinating corollary to the analysis: There exists a class of point-like defects which are ideal seeds for the growth of PPTs. If a cluster contains one of these defects, a dead surface is never obtained and tiles are forced throughout the plane. In terms of our stochastic growth analogy, a quasicrystal grain with one of these defects would always have sticky regions on its surface which promote perfect grain growth. The defects are a subclass of illegal Penrose tiling arrangements known as "decapods" [14]. One can assign a "charge" to a decapod defect using the Penrose arrow rules: Consider any closed path around the decapod running along tile edges. Add +1 if the arrow is along the path and −1 if the arrow is opposite the path. The net charge about any closed path in a PPT is zero. A dead surface can have net charge 0, ±2 only: (a) A dead surface edge has net charge zero; (b) all 108 degree corners have net charge zero; but (c) 72 degree corners can have charge ±1. However, a dead surface can have at most two 72 degree corners (e.g., a 72 degree rhombus). Decapod defects, though, can have charges ranging from -10 to 10. All decapod defects with charge greater than 2 in absolute value can never be enclosed by a dead surface. See Fig. 2.

If there is a decapod defect in the initial seed cluster, perfect growth continues ad infinitum with *absolutely no reduction in speed compared to the fastest crystal growth algorithm* using Rule 1 alone, attaching only at forced sites. Rapid growth is guaranteed for any finite p_f provided $p_c = p_o = 0$ — unarguably physically reasonable conditions.

Finally, what happens if the growth rules are violated at some isolated point during the growth process (e.g., p_f/p_c is finite)? Is the long-range order destroyed during subsequent growth? D. DiVincenzo, and, independently, M. Ronchetti and M. Jarić have been

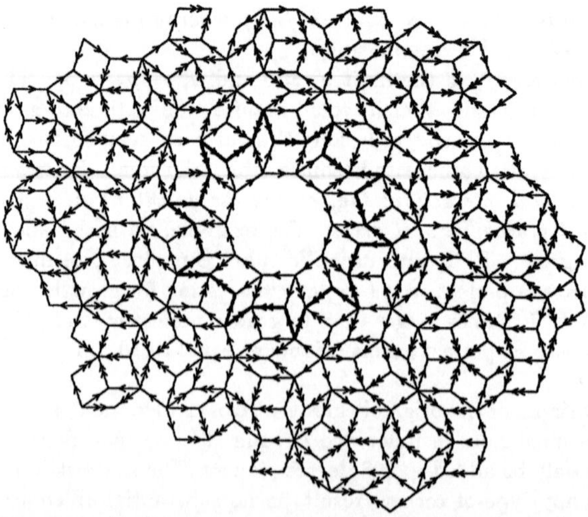

Figure 2: An illustration of growth beginning from a seed cluster with a charge-10 decapod defect. See text for definition of charge. The seed cluster is outlined in black. By adding tiles only at forced vertices, growth proceeds without ever reaching a dead surface.

exploring this issue recently. The approach has been to develop other local growth rules in case such errors appear. It appears at the present time that one can devise further local rules which restrict errors to point defects such that long range order is not destroyed.

4 Restorability, Growability, and the LI Class Problem

To reconstruct the atomic structure of a quasicrystal, it is necessary to determine its symmetry, the atomic decoration of its unit cells, and the unit cell packing arrangement. Whereas only one perfect packing (up to translations) is possible for periodic crystals, there are an uncountably-infinite number of "distinct," perfect quasicrystal packings — geometrically distinguishable arrangements of unit cells — for any given symmetry. The distinct packings are said to belong to different "local isomorphism" (LI) classes [15]. A key problem is that there is no method for determining both the correct LI class and the unit cell decoration simultaneously from diffraction data. The only distinguishing diffraction features are Bragg peak phases and intensities; however, in practice, it is not possible to separate effects due to changes in the unit cell decoration versus changes in the LI class packings of the unit cells. Most reported structural models employ *ad hoc* assumptions about the proper LI class, but this is clearly an unsatisfactory approach.

Recently, K. Ingersent and the author have considered two criteria that may be used to identify which LI classes are physically realizable in solids. The first criterion is that the packing should be *restorable* [16] — the packing can be uniquely specified given only a set of rules which fix the allowed unit cell clusters smaller than some bound. The restorability rules are longer-range generalizations of the Penrose edge-matching rules. The second criterion is that the packing should be *growable* in the sense defined above — the perfect packing may be grown from a seed cluster of unit cells using a set of local aggregation rules. The notion is that both sets of local rules might be manifested through finite-range atomic interactions. Restorability seems necessary if the LI class is to be the ground state of a Hamiltonian entailing only local interactions; growability seems necessary for the state to be physically realizable.

To be more specific, we have applied these criteria to the fivefold tilings obtained by projecting from five- to two-dimensions [17]. In choosing the projection plane, there are three degrees of freedom. Two correspond to phason translations of the quasicrystal, but the third — the projection along the (11111) axis — corresponds to changes in LI class. If we call γ the projection along the (11111) axis, then $\gamma = 0$ mod 1 corresponds to the Penrose tiling. Shifts to gamma between 0 and 1 produce a continuous spectrum of tilings belonging to other, non-Penrose LI classes.

It is conceivable that the Penrose LI class ($\gamma = 0$) turned out to be the unique LI class that is both restorable and growable. If so, one could argue that it is the only physically reasonable candidate LI class for describing a 2D pentagonal quasicrystal. An analogous result may exist for icosahedral symmetry. If this were the case, then determining the atomic structure would be reduced to finding the atomic decoration of the two tile types, as tractable as the crystal problem. However, our conclusion is not quite so simple, although perhaps more remarkable.

First, we have found restorability rules exist for an infinite subset of LI classes. To my knowledge, this is the first time that such rules have been identified for any LI class other than $\gamma = 0$, the original Penrose tiling. We find that we must generalize the notion from Penrose's nearest-neighbor edge-matching rule to a longer-range "cluster" rule that restricts allowed configurations of tiles out to some finite range, r. In this context, it is useful to introduce the notion of the "r-map," [16] the configuration of tiles centered at a given vertex which lies inside radius r. If two vertices have the exact same configuration of tiles (up to overall rotation by $2\pi/5$) surrounding them out to distance r, the vertices are said to have the same r-map. The complete collection of r-maps for all vertices in a given LI class is termed its "r-atlas." Restorability requires that there exists an r such that the r-atlas is unique to the given LI class. This extends the matching rule concept in a natural way. The physical notion is that, if we were to decorate the tiles with atoms, certain tile clusters would be of low energy (e.g., as discovered using a finite atom-cluster electronic structure calculation [18]); the set of low-energy clusters would determine the r-maps, and, thereby, the ground state structure. The restorability (or, equivalently, generalized matching) rules are equivalent to specifying the lowest energy clusters. Perhaps this way of viewing matching rules is more palatable physically and makes contact with more standard approaches for understanding atomic structure.

Even with restorability rules, though, there remains the issue of local growth rules. The key factor is establishing which LI classes have a "bounded context" inflation rule. An inflation rule is needed to prove that the set of dead surface shapes form a self-similar set so that, to establish that a corner rule, say, ensures growth from one dead surface to another, it is sufficient to check the smallest dead surfaces of each type. Here again, we generalize the notion of inflation rules as applied to the original Penrose tiling. In the original Penrose tiling, the inflation consists of a rule that subdivides each fat tile or each skinny tile in the same way. A weaker, but sufficient condition is that the way in which a given tile is subdivided can be determined by checking the configuration of tiles within some bounded region of the given tile. After developing a considerable technology to address this issue, our conclusion is that, of the continuous spectrum of LI classes defined by different choices of γ varying continuous between 0 and 1, only a discrete subset of measure zero is both restorable and growable:

$$\gamma = n\tau + m,$$

where $\tau = (1 + \sqrt{5})/2$ is the golden ratio and n and m are integers. See Fig. 3.

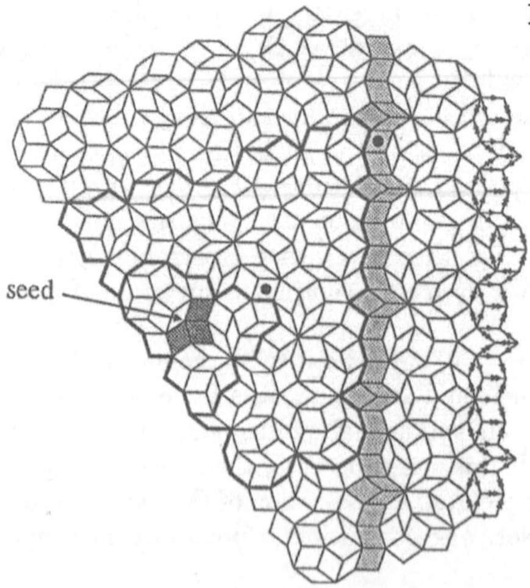

Figure 3: A cluster obtained during growth of a $\gamma = 2\tau$ tiling from a seed cluster (shaded), highlighting the dead surface encountered (heavy lines) and added corner tiles (black dot). Unlike the original Penrose tiling, there are two types of worm rows (shaded and with edge arrow marked) whose orientations are coupled (see Ref. 9).

Although the set is of measure zero, it is still infinite, and so it appears that we are still plagued with the LI class problem. However, there is the issue of the requisite interaction range. As it turns out, we find that the necessary range for both the restorability and growth rules scales $\propto |n|$. That is, matching and growth rules entail checking configurations of tiles beyond the nearest vertex by a distance of $\approx |n|$ tiles. We learn from this that the Penrose tile is special because it has the shortest-range matching and growth rules. Also, only a handful of tilings, say $|n| < 5$, have physically realistic interaction ranges. Hence, we can use the matching and growth rule criteria to reduce the LI class ambiguity from a continuous spectrum to a tractable handful of choices. If similar results can be obtained for icosahedral and other symmetries, it would represent a powerful application of the matching and growth rules that addresses a key issue in structural determination.

We note that, in the process of proof, we have discovered several mathematical properties of the tilings that may be useful in exploring other physical issues. We have identified generalized inflation rules, Ammann lines, "worm" configurations, and phason defects. These mathematical tools are presently being applied to studying which electronic and elastic properties are universal among the LI classes (and, hence, are determined by the long-range order of the structure), versus those properties which depend on the particular local topology of the given LI class. Previously, studies have been based on the original Penrose tiling only, and so it has not been possible to distinguish properties in this way.

5 Concluding Remarks

The principal conclusion to be drawn from this analysis of growth rules is that the intuition and arguments suggesting a fundamental impediment to growing perfect quasicrystals are absolutely wrong. In the analytical process, we have greatly expanded our catalog of tilings with local matching and growth rules. We have also addressed the LI class ambiguity problem, a key issue in the determination of quasicrystal atomic structure.

With hindsight, we can see that the strong intuition and arguments indicating that local growth rules are impossible did not take proper account of the role of orientational symmetry. The arguments are really based on the 1D case where local growth rules (and matching rules) are truly impossible to obtain. The 1D arguments can be extended to 2D and 3D incommensurate crystals, lattices with quasiperiodicity along two or three directions, respectively. However, the 2D pentagonal quasicrystal involves a grid of five sets of quasiperiodically spaced "rows" of tiles. Apparently, the sequence of rows to be added along any direction are *locally* constrained by the sequence of criss-crossing rows oriented in other directions. In a growing cluster with non-crystallographic symmetry, these constraints act to force the correct sequence in all five directions.

This suggests a valuable lesson to be learned from this study that may be important in addressing other physical properties. Most studies of growth, electronic, phonon, or structural properties have been inspired by the treatment of the 1D case. Here, though, we have seen that consideration of the noncrystalline nature of the orientational symmetry dramatically changes the result. Perhaps similar surprising breaks with 1D intuition must be sought in the studies of other physical properties if we are to uncover the most exciting, unique features of quasicrystals.

It is a pleasure to thank Prof. Fujiwara for his organization of a "perfect quasicrystal" Taniguchi Symposium. I also wish to thank Profs. Kubo, Takeuchi, Kamimura, and Nosé for their generous hospitality during my visit to Japan. This paper describes work done in collaboration with G. Onoda, D. DiVincenzo, J. Socolar and K. Ingersent. I wish to thank the Institute for Advanced Study in Princeton, NJ, for providing generous support for the completion of this project. The research was also supported in part by the National Science Foundation MRL Program under Grant No. DMR-8519059.

References

[1] D. Shechtman, I. Blech, D. Gratias, and J. W. Cahn, *Phys. Rev. Lett.* **53**, 1951 (1984).

[2] L. Pauling, *Nature* **317**, 512 (1985); *Phys. Rev. Lett.* **58**, 365 (1987); M. J. Carr, *J. Appl. Phys.* **59**, 1063 (1986).

[3] D. Shechtman and I. Blech, *Metall. Trans.* **16A**, 1005 (1985); P. W. Stephens and A. I. Goldman, *Phys. Rev. Lett.* **56**, 1168 (1986); **57**, 2331 (1986).

[4] D. Levine and P. J. Steinhardt, *Phys. Rev. Lett.* **53**, 2477 (1984).

[5] P. A. Horn, P. A. Bancel, P. M. Horn, J. L. Jordan, S. LaPlaca, J. Angilello and F. W. Gayle, *Science* **238**, 660 (1987).

[6] P. J. Steinhardt, *Science* **238**, 1242 (1987).

[7] G. Y. Onoda, P. J. Steinhardt, D. P. DiVincenzo, and J.E.S. Socolar, *Phys. Rev. Lett.* **60**, 2653 (1988).

[8] For *AlRuCu*, see C. A. Guryan, A. I. Goldman, P. W. Stephens, K. Hiraga, A. P. Tsai, A. Inoue, and T. Masumoto, *Phys. Rev. Lett.* **62**, 2409 (1989); For *AlFeCu*, see P. Bancel, IBM Preprint (1989).

[9] K. Ingersent and P. J. Steinhardt, "Matching Rules and Growth Rules for Pentagonal Quasicrystal Tilings," Penn Preprint (1989).

[10] R. Penrose *Bull. Inst. Math. and Its Appl.* **10**, 266 (1974).

[11] N. de Bruijn, *Ned. Akad. Weten. Proc. Ser. A* **43** 39,53 (1981).

[12] Note that this rule differs slightly from the rule listed in Reference 7. In that reference, fat rhombuses are only added to 108 degree corners. Either rule can be used. See later discussion.

[13] M. Ronchetti and M. Jarić, *Phys. Rev. Lett.* **62**, 1209 (1989); G. Y. Onoda, *et al.*, *op. cit.*, 1210 (1989).

[14] M. Gardner, *Sci. Am.* **236**, 110 (1977).

[15] For a discussion of the Local Isomorphism Class Problem, see D. Levine and P. J. Steinhardt, *Phys. Rev. B***34**, 596 (1986); also, Chaps. 1 and 3 in *The Physics of Quasicrystals*, ed. by P. J. Steinhardt and S. Ostlund, (World Scientific, Singapore, 1987).

[16] The notion of "restorability" is first discussed by L. S. Levitov, *Comm. Math. Phys.* **119**, 627 (1988).

[17] A. Katz and M. Duneau, *J. de Physique* **47**, 191 (1986); V. Elser, *Acta Cryst. A***42**, 36 (1986).

[18] See, for example, the discussion of electronic structure calculations by T. Fujiwara in this volume.

The Alternation Condition and 2D Quasicrystals

J.E.S. Socolar

Physics Department, Harvard University, Cambridge, MA 02138, USA

Abstract. The alternation condition can be applied to the rhombic tiles required for constructing N-fold symmetric 2D space-filling tilings and can be enforced via a local rule specifying allowed tile configurations. For N not divisible by four, any tiling that satisfies the alternation condition can differ from a perfectly ordered quasicrystal by at most a set of uniformly bounded fluctuations in "perp-space". Among tilings that satisfy the alternation condition, Penrose tilings are special in two interesting ways.

1 Introduction

Recent experiments on the decagonal phase of $Al_{65}Co_{20}Cu_{15}$ suggest that it can be described as a decoration of a quasicrystal tiling with 10-fold symmetry, repeated periodically along its 10-fold axis. [1] If further investigations demonstrate conclusively that the phase is an ordered quasicrystal (i.e., that the phason fluctuations expected in other models for the structure are not present), then a complete understanding of the structure will have to include an identification of the local interactions responsible for propagating the order. Such interactions can be represented by matching rules for decorated tiles. Unfortunately, while several examples of tilings with matching rules are available, including tilings with 5- and 10-fold symmetry (the original Penrose tilings [2,3]), 8- and 12-fold symmetry [4] and icosahedral symmetry [5,6], it is not clear which features of the rules are essential for the establishment of quasicrystal order. Thus we do not know how to tell whether matching rules are being implemented unless the system happens to be described in detail by one of the known examples. A more general theory of matching rules is needed, as well as a more profound understanding of the existing examples.

Several insights can be gained from the analysis of the "alternation condition" (AC), a simple rule (defined below) that can be applied to 2D rhombic tilings of any symmetry. It is found that the Penrose matching rules are a special case of a much more general class which includes rules for tilings of N-fold symmetry. For any N not divisible by four, all tilings satisfying the AC produce Bragg diffraction at the positions of a qua-

Springer Series in Solid-State Sciences, Vol. 93 Quasicrystals
Editors: T. Fujiwara · T. Ogawa

sicrystal reciprocal lattice along with some diffuse scattering, the effect of the AC being to enforce a bound on "perp-space" fluctuations.

In all previously known examples of tilings with matching rules the incommensurate ratios associated with their symmetries are quadratic, as has been emphasized by Levitov in Ref. [7]. There it is shown that for no other symmetries do the tilings produced by projection from a hypercubic lattice support perfect matching rules; i.e., local rules that restrict a tiling to a particular local isomorphism class and hence eliminate all diffuse scattering. Nevertheless, "weak matching rules", which permit fluctuations that generate diffuse scattering but still guarantee the existence of Bragg peaks, do exist for other symmetries and the AC provides a basis for their construction. [8] This surprising result sheds new light on the geometrical origin of matching rules.

In this paper, the theory of tilings satisfying the AC is sketched with the intent of presenting concepts and results rather than rigorous justifications. After some preliminary definitions are given, the weak matching rules generated by the AC are discussed, the features that distinguish the Penrose tilings are considered, and the case of N divisible by four is discussed. Although the discussion in this paper is limited to 2D tilings, there is a natural extension of the AC to rhombohedral tilings of icosahedral symmetry, details of which are given in Ref. [8].

2 The alternation condition

It is assumed that the reader is familiar with the generalized dual method for generating a rhombic tiling from a set of intersecting grids and the canonical method of generating rhombic tilings by projection from a hypercubic lattice. (See, for example, articles in Ref. [9].) For the purposes of this paper it is more useful to refer to the generalized dual method than the projection technique, but there is a direct relation between the two. [10] Any rhombic tiling that is generated by projection from a hypercubic lattice in the usual way is the dual of a periodic N-grid. Such a tiling will be called a "canonical projection".

To establish notation we review the definitions of some of the elements of the generalized dual method. For now, we will only be interested in N-fold symmetries where $N = Q$ or $2Q$ and Q is an odd number. We will return to the case of N divisible by four in Section 4. A "Q-fold star" is the set of unit vectors $\mathbf{e}_n \equiv (\cos(2\pi n/Q), \sin(2\pi n/Q))$, where $n = 0, \ldots, Q-1$. A "periodic grid" is a set of equally spaced parallel lines normal to a "star vector" $\mathbf{e}$. A "periodic Q-grid" is a network of Q intersecting periodic grids, no two of which have parallel star vectors. The "phase" of each grid is denoted by γ_n, an integral value of the phase meaning that one line of the grid passes through the origin.

A Q-fold symmetric tiling can be obtained by taking the dual of a periodic Q-grid with the star vectors of a Q-fold star. It can be shown that Q-fold symmetry requires certain constraints on the phases: if there exists a subset of star vectors for which $\sum_i \mathbf{e}_i = 0$, then one must have $\sum_i \gamma_i = \sum_i \gamma_{i+j}$ for all j, where all subscripts are taken modulo Q. For $N = 2Q$, an N-fold symmetric tiling can be obtained by taking the dual to a Q-grid based on a Q-fold star if an additional constraint is met: the sum of all Q phases must be either an integer or half an integer. (For example, the Penrose tiling has 10-fold symmetry and is the dual of a periodic 5-grid with the sum of the phases equal to an integer. [3])

Each open region in a periodic Q-grid corresponds to a vertex v in the dual tiling and is defined by a set of integers k_n, where the region in question lies between the lines of index k_n and $k_n + 1$ in the grid normal to $\mathbf{e}_n$. With each vertex we associate the quantity $\Gamma(v) = \sum_{n=0}^{Q-1} k_n$ and the vectors $\mathbf{w}^{(i)} = \sum_{n=0}^{Q-1} k_n \mathbf{e}_{n \times i}$, where $i = 1, \ldots, (Q-1)/2$. Note that $\mathbf{w}^{(1)}(v)$ is the position of the vertex in the tiling plane. The other $\mathbf{w}^{(i)}$ together with Γ can be thought of as positions of the vertex in subspaces orthogonal to the tiling plane in the space of the hypercubic lattice of the projection technique; i.e., Γ and $\mathbf{w}^{(i>1)}$ are the perp-space variables associated with the tiling. Note that the four vertices dual to the regions surrounding a single intersection in a Q-grid form a rhombic tile.

A row is the set of tiles dual to the intersections of a single grid line with all of the other grids. The row consists of a continuous string of tiles joined along edges parallel to the star vector of the grid line, as shown in Figure 1a. In a given row each tile shape can occur in two orientations that are related by reflection through a line parallel to the star vector that defines the row. (This will not be true for N-fold symmetries with N divisible by four, where square tiles appear.)

The AC specifies that for every tile shape and every row in the tiling the orientation of the shape must alternate along the row. That is, on any row, between any two tiles of the same shape and orientation there must be a tile of the same shape but opposite orientation. Note that the distance between a tile and the next one of its shape can be large. The row in Figure 1a satisfies the AC. Tiles of like color all have the same shape and their orientations alternate.

It is easy to see that any canonical projection satisfies the AC. In any row dual to a grid line L, tiles of the same shape but opposite orientations are dual to intersections of L with grids that intersect L at supplementary angles. (See Figure 1b.) For periodic grids with equal spacings, the spacing between intersections of one orientation is the same as the spacing between the others and hence the two must alternate. Now canonical projections are not the only tilings satisfying the AC.

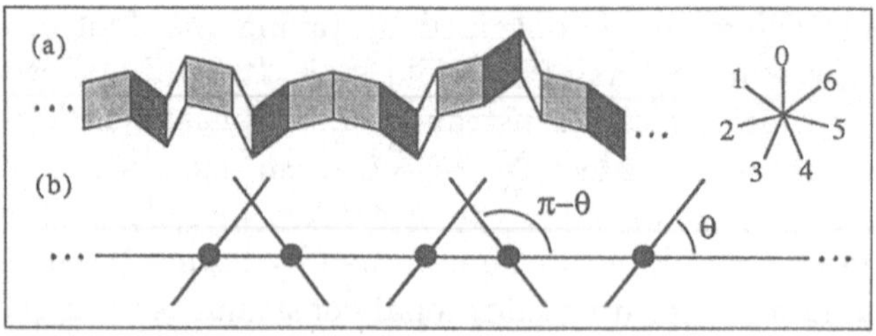

Figure 1: (a) A row of tiles dual to a grid line normal to $\mathbf{e}_0$. The star vectors are shown at right. (b) Intersections dual to tiles of the same shape in a single row. Only the grid lines corresponding to the black tiles in (a) are shown.

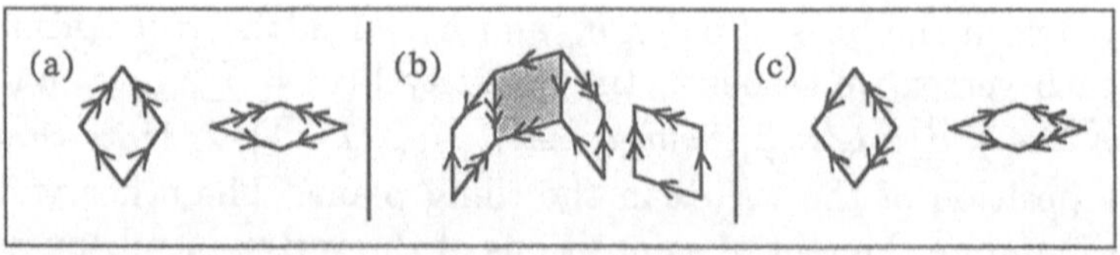

Figure 2: Decorated tiles for 5-fold symmetry. (a) The Penrose tiles. (b) A row of Penrose tiles demonstrating that the decorations enforce the AC. (c) The tiles that can be added to the Penrose set without allowing violations of the AC.

Nevertheless, it can be shown that: [8]

1. For any N not divisible by four, the AC can be implemented as a rule specifying the allowed configurations of tiles within a finite neighborhood.
2. *For P-fold symmetric tiles, where P is a prime number, the AC is sufficient to guarantee that deviations from a perfect quasicrystal in the perp-space variables are uniformly bounded (in an infinite space-filling tiling with no overlapping tiles).*
3. When Q is odd but not prime, the AC together with another local rule requiring that every orientation of every tile shape must occur within every circle of some specified radius ensure that deviations from a perfect quasicrystal in the perp-space variables are uniformly bounded.

Tilings with N-fold symmetry where $N = 2Q$ are just special cases of tilings with Q-fold symmetry and are covered by these results as well.

An examination of the Penrose matching rules (Figure 2a) immediately reveals that they enforce the AC. Consider the tiles shown in Figure 2b. Suppose we are to build a row by adding to the shaded fat tile along its right side. It is clear that if we were to add a fat tile, matching the single arrows would require that the new tile be the mirror image of

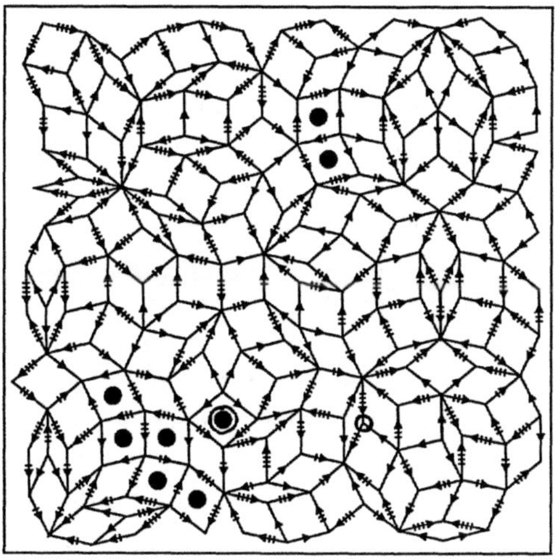

Figure 3: A 7-fold tiling composed of tiles that enforce the AC. The black dots mark eight distinct decorations of one tile shape. The circled black dot marks the tile decoration that does not appear in canonical projections when $\sum \gamma_n$ is an integer and corresponds to the tile at the origin in Figure 4. The tiles sharing the vertex marked by an open circle form a transformable hexagon. (See Section 3.)

the shaded tile. If we add a skinny tile, as shown in the figure, and then add a fat tile to that, we find that matching the double arrows again requires that we add the mirror image of the shaded tile. No matter how many skinny tiles are interposed, in fact, the next fat tile to be added must be the mirror image of the shaded tile. Similar analysis of the skinny tiles reveals that they too must alternate in orientation.

A closer look at the way in which the AC is implemented by the Penrose matching rules shows that the essential features are that parallel edges of a single tile have different arrow types and that on the fat tile arrows on opposite edges have the same direction, while on the skinny tile they have opposite directions. One sees that the two tiles of Figure 2c can be added to the Penrose set and the AC will still be enforced. The four tile set is needed to construct 5-fold canonical projections that are not in the Penrose local isomorphism class, as first noted by Kleman and Pavlovitch. [11]

Generalization to higher symmetries of the construction of the decorated tiles that enforce the AC is straightforward, though somewhat tedious to describe in detail. The case of 7-fold symmetry is illustrated in Figure 3. Here four types of arrow are necessary and there are eight distinct decorations of each shape. To see why four types of arrows are needed, imagine that we have completed half of a row and consider what must be specified by the arrow on the edge where we are about to add. There are three tile shapes and the last appearance of each one in the row could have had either of the two possible orientations. Thus the arrow must distinguish eight possible situations. This can be accomplished with four types of arrow, each of which has two possible orientations.

Counting the distinct tile decorations is a bit more complicated: Opposite edges of each tile of shape S are required to have arrows that specify identical orientations for tiles of shapes other than S and opposite orientations for shape S. In this way, as a tile is traversed from one edge to the opposite edge the decoration keeps track of the expected orientation of the next tile in the row according to the AC. Now once a given edge of a tile is decorated the decoration on the opposite edge is determined. In addition one bit of information on each edge is fixed by the the orientation of the tile itself. Thus it appears that each of two adjacent edges of the tile must be decorated in one of four ways, making 16 possibilities. The tile decorations have been double counted, however, as they come in pairs related by rotation of the tile by 180°.

In the case of Q-fold symmetry (or $2Q$-fold symmetry), one finds that there are $S = (Q-1)/2$ tile shapes, $Y = 2^{S-1} = 2^{(Q-3)/2}$ arrow types, and $Y^2/2 = 2^{Q-4}$ distinct decorated tiles of a given shape (counting enantiomorphs as distinct).

It is instructive to examine the proof that the AC is a weak matching rule for tilings of P-fold symmetry where P is a prime. The proof given in Ref. [8] is a reductio ad absurdum in which it is shown that a large difference in the values of a perp-space variable at any two vertices in the tiling inevitably leads to a geometric inconsistency: for certain sequences of vertices v_1 and v_2 for which $\|\mathbf{w}^{(1)}(v_1) - \mathbf{w}^{(1)}(v_2)\|$ is always bounded, the AC forces a divergence in $\|\mathbf{w}^{(i)}(v_1) - \mathbf{w}^{(i)}(v_2)\|$ for $i > 1$, but this is impossible since v_1 and v_2 must be joined by a bounded number of tile edges if the tiling is space-filling and does not contain overlapping tiles. To avoid the contradiction, one must conclude that the assumed large difference in perp-space never occurs. Note that the degree of irrationality of $\cos(2\pi/Q)$ never enters the proof. *The existence of weak matching rules does not depend on* $\cos(2\pi/Q)$ *being a quadratic irrational.*

In analyzing the proof one also sees that the AC is unnecessarily restrictive. What is really necessary is a rule specifying that for any continuous segment of any row the difference between the numbers of tiles of the same shape but opposite orientations is bounded. Thus, for example, one could allow two consecutive tiles of the same orientation if the next two tiles of that same shape were guaranteed to have the other orientation. Admitting such possibilities, however, quickly leads to much more complicated tile decoration schemes for nearest-neighbor rules.

3 Another look at the Penrose tiles

At no point in the proof that the AC generates weak matching rules do any unique properties of the Penrose tilings appear. There are, however, two features that distinguish the Penrose tiles of Figure 2a from other sets of decorated tiles constructed to enforce the AC.

The first special property applies to the entire set of 5-fold tiles (the four tiles shown in Figures 2a and 2c), distinguishing this set from those of higher Q-fold symmetries. For $Q \geq 7$ one can always construct irregular hexagons from three tiles of different shapes. (See Figure 3.) These are called "transformable" hexagons because the three tiles can be rearranged within the hexagon boundary without affecting the AC on any row. (The rearrangement corresponds to rotating the hexagon through 180°.) No tile changes its orientation upon rearrangement. On each of the three rows passing through the hexagon two tiles of different shape exchange positions, but this has no effect on the order of tiles of the same shape in any row. In contrast, for a hexagon containing two tiles of the same shape, which must be of different orientations, only one of the arrangements can satisfy the AC since rearrangement reverses the order of the two like shapes in the row containing them and only one choice is consistent with the other tiles in that row. Now in the case of 5-fold symmetry there are only two tile shapes. Thus every hexagon must contain two tiles of the same shape and there are no transformable hexagons.

The presence of transformable hexagons in a tiling obeying the AC immediately implies that the matching rules are at best weak. Since transformable hexagons can be flipped at random throughout the tiling, it is clear that the generic tiling will generate diffuse scattering in addition to its Bragg peaks. By the same token, the absence of transformable hexagons in the 5-fold tilings suggests that the AC is a stronger rule in that case. Although it has not been rigorously proven, it appears that every 5-fold tiling satisfying the AC is in fact a canonical projection, up to the possible transformation of one entire row of hexagons. (Note that if an entire row consists of pairs of tiles that lie in hexagons and if all of the hexagons are simultaneously transformed, then the AC is not affected.)

The second feature that distinguishes the Penrose tiles is simply the fact that they are only a subset of the full 5-fold set. The effect of this is to reduce the set of tilings composed of the Penrose tiles to a single local isomorphism class, a property that has significant physical implications. [12] It is interesting to note that for any P-fold symmetry (P is prime) there is a particular local isomorphism class of canonical projections in which one type of decorated tile of each shape does not appear. [13] This is proven via consideration of the phases of the P-grids, which are constrained to satisfy certain inequalities in order for a given tile decoration to appear in the dual tiling.

Figure 4 shows how the constraints arise in the 7-fold case for the tile with acute angle $3\pi/7$. We sketch the proof for this case only. The gen-

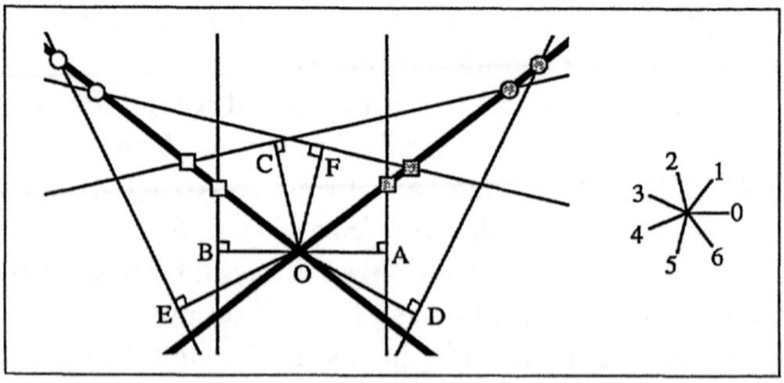

Figure 4: Construction used in the proof of the nonexistence of certain tile decorations when $\sum \gamma_n$ is an integer. Two grid lines normal to $\mathbf{e}_0$ are shown and one normal to each of the other vectors in the 7-fold star shown at right. In directions $2, \ldots, 5$, the grid lines passing closest to O but not below it are shown. Intersections marked with the same symbol correspond to tiles of the same shape and opposite orientation along the rows dual to the heavy grid lines. With O defined as the origin, we have OA$= \gamma_0$; OB$= 1 - \gamma_0$; OC$= \gamma_2$; OD$= 1 - \gamma_3$; OE$= \gamma_4$; and OF$= 1 - \gamma_5$.

eralization to other tile shapes and other symmetries is straightforward. Choose the origin in the 7-grid to lie at the intersection of lines normal to $\mathbf{e}_1$ and $\mathbf{e}_6$ (the heavy lines in the figure). The tile dual to the intersection at the origin has the shape we are interested in. The decoration of this tile is determined by the order in which the marked intersections in the figure occur. To achieve the decoration corresponding to the arrangement shown in the figure, the following inequalities must be satisfied:

$$\gamma_0 < 1 - \gamma_5; \quad 1 - \gamma_0 < \gamma_2; \quad \gamma_2 < 1 - \gamma_3; \quad 1 - \gamma_5 < \gamma_4.$$

By construction, we have $\gamma_1 = \gamma_6 = 0$ and $0 < \gamma_n < 1$ for all other n. Let $\Gamma = \sum_{n=0}^{6} \gamma_n$. (Note that only the fractional part of Γ is meaningful due to the freedom to choose which grid line has index 0.) The equation above implies $\gamma_3 < \Gamma - 1 < \gamma_4$. For general Γ this can indeed occur, but for Γ equal to an integer it clearly cannot. Symmetry and the freedom to choose the origin then imply that any canonical projection for which Γ is an integer contains no tiles with the decoration corresponding to Figure 4.

It is of course well known that the Penrose tilings are dual to 5-grids for which the sum of the phases is an integer. [3] It is not known whether the analogous tilings dual to P-grids with $P \geq 7$ share any special properties other than that the P-grids for which the sum is an integer yield tilings with $2P$-fold symmetry.

4 When N is divisible by 4

It seems odd that N-fold symmetries with N divisible by four should be singled out with regard to their matching rule properties. The mere presence of square tiles does not seem like it should prohibit the existence of matching rules and there do not appear to be any other striking differences between . One might even expect that the AC would still generate weak matching rules when applied to the non-square tiles only. It is easy, however, to construct a tiling with $N = 4M$-fold symmetry (for any integer $M \geq 2$) that obeys this form of the AC but is not bounded in perp-space. The construction makes transparent the important difference between $N = 4M$ and other symmetries and indicates what sort of supplement to the AC may be needed to generate weak matching rules.

Let $N = 2N'$ with N' an even number. To make an N-fold symmetric tiling we use a N'-grid with star vectors $\mathbf{e}_n = (\cos(2n\pi/N), \sin(2n\pi/N))$ where $n = 0, \ldots, N'-1$. (Note that a complete N-fold star has inversion symmetry so that all the needed directions are included in half of the star.) Now consider an N'-grid in which each grid is periodic but the spacing between lines has one value for all even n and a different value for all odd n. The dual to such an N'-grid cannot be bounded in perp-space, one indication being that the symmetry between the number of $\mathbf{e}_n$ and $\mathbf{e}_m$ edges is broken when $n - m$ is odd. On the other hand, it is easy to see that the tiling satisfies the AC. For a grid line normal to $\mathbf{e}_n$, tiles of the same shape and opposite orientation are dual to intersections with grids normal to $\mathbf{e}_{n+i}$ and $\mathbf{e}_{n-i}$, where subscripts are taken $\mathrm{mod}_{N'}$ and i is any integer. Since N' is even, $\mathrm{mod}_{N'}(n \pm i)$ are either both odd or both even, so by construction the two grids have the same spacing and the intersections of them with the original line must alternate.

Two points are worth noting here: First, the duals to the grids we have constructed can be shown to differ from a perfect quasicrystal only by a uniform phason strain. Furthermore, while the magnitude of the strain is arbitrary, its symmetry is unique. All other uniform phason strains require violations of the AC. It can be shown that for any N divisible by four there exist two phason strains that transform under 1D irreducible representations of the group C_{Nv} and that above configuration corresponds to one of these. Since fluctuations in the magnitude of a 1D strain must induce fluctuations in other strain components, the following conjecture is suggested: The AC applied to N-fold symmetric tilings with N divisible by four allows an average gradient in perp-space but restricts fluctuations about this average to be bounded, ensuring the existence of Bragg peaks, though they may not be positioned at the ideal quasicrystal reciprocal lattice points.

Second, a comparison with the known perfect matching rules for the 8- and 12-fold cases is revealing. In those cases the existence of a phason strain that does not violate the AC has already been noted. The AC is enforced by the "edge-arrow" decorations described in Ref. [4], where explicit examples are given of tilings with uniform phason strains in which the edge-arrows always match properly. The edge-arrow rules are also shown to correspond to the continuity of certain lines (Ammann lines) that run parallel to the grid lines and can be produced by a simple decoration of the tiling. The construction of perfect matching rules is accomplished by adding a rule that enforces the continuity of an additional set of lines that run normal to the $\mathbf{e}_n + \mathbf{e}_{n+1}$ directions (the secondary Ammann lines) . Thus we are led to conjecture that some method of supplementing the AC to produce weak matching rules exists in the general case and that this method will involve rules that constrain configurations on "secondary rows" of tiles that run in these additional directions.

5 Conclusion

The physical relevance of weak matching rules is not entirely clear. It is not known, for example, whether it is possible for local growth algorithms to generate tilings obeying the AC for any symmetries other than $N = 5, 8, 10$ or 12, as may be necessary for the ordered quasicrystal state to be accessible in a real system. Furthermore, weak matching rules cannot be a completely accurate description of a system at $T = 0$, where differences between energies of tilings in different local isomorphism classes must select a single class, or perhaps favor a crystalline phase. Nevertheless, the discovery that the AC generates weak matching rules provides a new framework for understanding matching rules in general and shows that a simply defined local rule can induce long-range quasicrystalline order.

Acknowledgements. It is a pleasure to thank D. DiVincenzo, K. Ingersent, A. Katz, L. Levitov, P. Steinhardt and L. Tang for sharing ideas relevant to the results presented here. Support for this work was provided by the Harvard Society of Fellows and the NSF through grant No. DMR-88-17291.

References

[1] A. R. Kortan, R. S. Becker, F. A. Thiel, and H. S. Chen. *AT&T Bell Laboratories preprint.*, 1989.

[2] R. Penrose. *Bull. Inst. Math. Appl.*, **10**:266, 1974.

[3] N. de Bruijn. *Proc. Kon. Nederl. Akad. Wetensch.*, A**84**:39, 1981.

[4] J. E. S. Socolar. *Phys. Rev.*, B**39**:10519, 1989.

[5] A. Katz. *Commun. Math. Phys.*, **118**:263, 1988.

[6] J. E. S. Socolar and P. J. Steinhardt. *Phys. Rev.*, B**34**:617, 1986.

[7] L. S. Levitov. *Commun. Math. Phys.*, **119**:627, 1988.

[8] J. E. S. Socolar. *Commun. Math. Phys.*, to appear, 1990.

[9] P. J. Steinhardt and S. Ostlund. **The Physics of Quasicrystals**. World Scientific, Singapore, 1987.

[10] F. Gahler and J. Rhyner. *J. Math. Phys.*, A**19**:267, 1986.

[11] M. Kleman and A. Pavlovitch. *J. Phys.*, **C3**, Workshop on Aperiodic Crystals:229, 1986.

[12] D. Levine and P. J. Steinhardt. *Phys. Rev.*, B**34**:596, 1986.

[13] When Q has prime factors, more than one tile decoration of each shape can be eliminated.

Random Tiling Model of Quasicrystalline Order

M. Widom

Department of Physics, Carnegie-Mellon University,
Pittsburgh, PA 15213, USA

Abstract. This paper reviews the theoretical and experimental evidence concerning the origins of thermodynamic stability of quasicrystalline structures. Random tiling models provide a compelling model of long range quasiperiodic order arising from tiling entropy. Theoretical predictions of this model are examined and contrasted with the quasiperiodic crystal and icosahedral glass models. Several experiments have been performed capable, in principle, of testing these models. The results have included a few surprises and in general tend to support the theory of random tilings.

1. Introduction

Shortly after the discovery by Shechtman, et al. [1] of icosahedral order in rapidly quenched metal alloys, theories of the nature and origin of the non-crystalline structures began to be proposed. Chief among these are the quasiperiodic crystal models [2], and the icosahedral glass models [3]. For reasons to be discussed, neither model proves entirely satisfactory as a description of real, thermodynamically stable, quasicrystalline alloys. Rather, there is a third type of model [4-6], based on random tilings, which shows great promise.

The three models differ both qualitatively and quantitatively in ways which are experimentally testable. Crudely speaking, the quasiperiodic crystal model makes sense when the quasicrystal forms the ground state of a system. The random tiling model is expected to apply when the quasicrystal is an equilibrium state at intermediate temperatures but not absolute zero. The icosahedral glass model is expected to apply when the quasicrystal phase is metastable. Many types of experimental tests of these models are possible. Scattering studies offer perhaps the most decisive tests. Thermodynamic measurements and microscopy are other types of experiments that could distinguish between the models as well. The remainder of this introduction describes the random tiling model. Then theoretical comparisons are made with the icosahedral glass and quasiperiodic crystal models in the following section. Section 3 discusses experimental tests which can distinguish between the models.

The random tiling models of quasicrystals are inspired in part by microscopic models of both real quasicrystal forming materials [7] and by theoretical systems which have been shown to have a quasicrystalline equilibrium state [8-9]. Let us consider this theoretical model. Figure 1a shows a pair of rhombic tiles. On each vertex of the tiles are placed atoms, either large or small. One can consider a few versions of this model with varying degrees of complexity. The most realistic version of this model ignores the rhombic tiles and allows long range interactions between the atoms, which move in the continuum. It may be observed, however, that low temperature states (figure 2a) are always sufficiently well ordered that the atoms form a well defined tiling (figure 2b). A cruder version of this model replaces the atoms with rigid tiles, and truncates the bonds at nearest neighbors.

These models are distinguished by a large entropy associated with the variety of ways the tiles can fill the plane. In the simplest truncated rigid tiling model the entropy re-

Springer Series in Solid-State Sciences, Vol. 93 **Quasicrystals**
Editors: T. Fujiwara · T. Ogawa

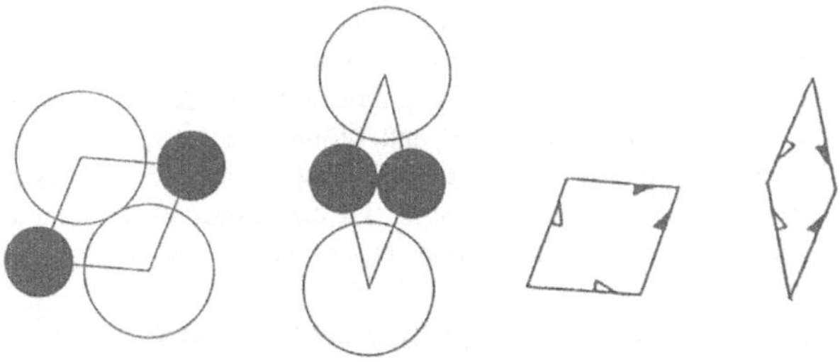

Figure 1. Large and small atoms decorate vertices of fat (72°) and thin (36°) rhombi (a). Arrows on tile edges must match to form Penrose tiling (b).

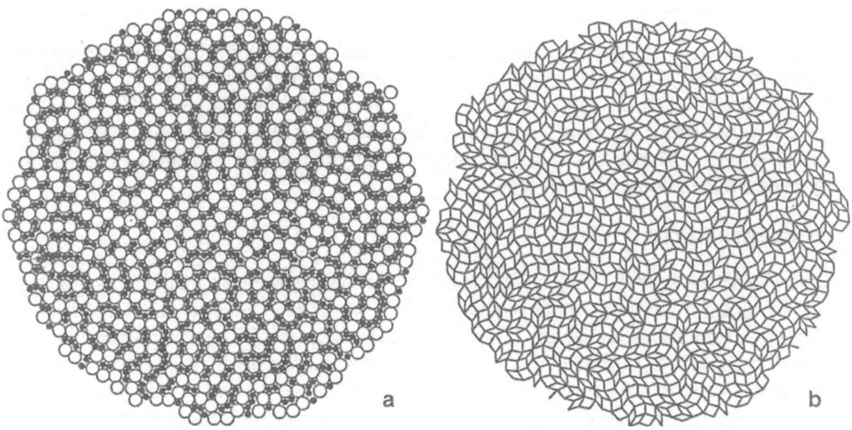

Figure 2. Atomic system in equilibrium at low temperature (a). Nearest neighbor large-small bonds define rhombus tiling (b).

flects an exact degeneracy. But even when the degeneracy is not exact the entropy can be large and significant. The entropy is made manifest in computer simulations [8,9] in which the structure is constantly fluctuating from one arrangement of atoms to another.

It is helpful to describe this model in the higher dimensional space projection language. Any rigid tiling of space can be thought of as a set of points in a fictitious space of finite but high dimensionality projected down into a lower dimensional physical space. Tilings of the plane by the rhombi shown in figure 1a correspond to a two dimensional surface of points in a five dimensional hypercubic lattice. Fluctuations between degenerate structures in the tiling correspond to fluctuations of this surface, which are also known as phasons. Local rearrangements of the tiles create a localized bump on the surface. Long wavelength phasons, which correspond to periodic undulations of the surface, require coherent rearrangement of tiles of large distances.

Defining the manner in which the system is quasicrystalline is somewhat problematic since the ordering is not perfect and there is no unique structure at all. Henley [4] suggested that if the entropy density of the tiling were to have a quadratic extremum as a function of phason strain, with a maximum at zero strain, the tiling would have quasiperiodic long range order in three dimensions but only quasi-long range order in two dimensions. In the projection language, perpendicular space heights are bounded in three dimensions but diverge logarithmically in two dimensions. Subsequent transfer matrix [5] and Monte Carlo [6] calculations have confirmed that for short truncated rigid tiling models all these assertions are correct. The entropy density has a quadratic maximum at zero, and the curvatures of the entropy yield phason elastic constants. That follows because the phason part of the elastic free energy is given by

$$F = -TS.$$

Thus the curvatures of entropy density multiplied by temperature are phason elastic constants.

2. Theoretical Models

This section contrasts the random tiling model with the icosahedral glass and the quasiperiodic crystal models. These two models stand at opposite extremes from one another, with the random tiling model in some sense a compromise between the two. The icosahedral glass model offers hope of describing metastable quasicrystalline structures. The quasiperiodic crystal model presents the most exciting new physics of these models, but it appears unlikely to describe real materials correctly.

2.1 Icosahedral glass models

The name "icosahedral glass" was introduced to describe a particular model of quasicrystalline order [3] but, for the purposes of this paper, I will use it as a name for a family of models. The unifying feature of these models is that they seek not to describe a precise atomic structure in thermodynamic equilibrium, but rather a growth process which could lead to structures with diffraction patterns and certain physical properties similar to real quasicrystalline materials. Because the models are defined by nonequilibrium growth processes, it is unreasonable to expect them to describe stable structures such as AlCuFe. Rather, these models create metastable states and are more appropriately applied to growth out of equilibrium.

Broadly speaking these models create a quasicrystal by randomly aggregating icosahedral or other shaped clusters [10] along axes of high symmetry. The clusters may be thought of as groups of atoms which clump together in the liquid phase. Or else one may consider them to be individual atoms for which the equilibrium structure would have a large degree of icosahedral order so that the icosahedral cluster formation is a part of the aggregation process. The most important success of this model is it showed how one could obtain long range orientational order and fairly sharp peaks without requiring long range translational order. By invoking the Hendricks - Teller mechanism it was understood why many diffraction peaks appeared superficially to be sharp even though the range of translational order was limited. In the projection picture of quasicrystals, the icosahedral glass model corresponds to a wildly fluctuating and torn surface. The fluctuations provide entropy, but the tears raise the energy to a point where the model is unlikely to be in equilibrium.

Much attention has focussed on the dependence of the peak width on phason momentum [11]. Icosahedral glass models tend to show peak widths depending quadratically on phason momentum $q_{\perp}$. Models in which there is basically a well ordered quasicrystal structure with frozen in dislocations and phason strains, in contrast, show widths growing linearly in $q_{\perp}$. Experiments on AlMn have tended to find a linear dependence at least for $q_{\perp}$ not too large. But this should not be taken as discrediting the icosahedral glass scenario, because there is surely some improved growth rule for which the proper dependence can be obtained over the range of $q_\perp$ which is measured. That assertion is, in a sense, tautology since it is certain that the quasicrystal state of AlMn is metastable and thus obtainable through some nonequilibrium growth process.

It is reasonable to put into the category of icosahedral glass models the recent growth algorithms devised by Onoda, et al. [12]. These algorithms generate almost perfect quasicrystalline order via local growth rules. They naturally fall into the icosahedral glass category because they grow the structure out of equilibrium. There are no rearrangements allowed once a tile is attached. The existence of such rules clearly shows that icosahedral glass models can cover a wide spectrum of degrees of disorder.

2.2 Quasiperiodic crystal models

The Quasiperiodic crystal model [2] is the most prominent model of quasicrystalline order. I will argue shortly that it is unlikely to be correct in certain important respects. But, because this model captures the essence of the structure of thermodynamically stable quasicrystals, and provides an especially simple description of atomic locations in infinite structures, the model is probably a sensible starting point for calculating physical properties of quasicrystalline materials.

In general the quasiperiodic crystal model states that the arrangement of atoms is similar in nature to that of a crystal - a superposition of a small number of periodic mass density waves - but that the periods may be incommensurate and the symmetry noncrystallographic [13]. In ordinary crystals the existence of an underlying crystal lattice leads to the recognition of a unit cell. The entire crystal structure may be described by a tiling of space with the unit cell once the atomic decoration of a single unit cell is found. Quasiperiodic crystals may be similarly described, but the space to be tiled is of a higher dimension than physical space.

In physical space it is natural to consider tiling models of quasicrystals such as the Penrose tiling model which fills the plane with two packing units. But as soon as we try to pack these units in physical space we are faced with a problem tiling quasiperiodically. There must be rules which guarantee that our tiling will be quasiperiodic. Such rules were found by Penrose and are shown in figure 1b by arrows along the edges of the tiles. The result of tiling the plane by these rules is a tiling which is every bit as ordered as an ordinary crystal, but is quasiperiodic instead of periodic. The diffraction pattern of such a structure contains delta function peaks and no diffuse scattering [2].

An implicit assumption in such a model is that the tiling rules are implemented in real materials by interactions among the atoms which decorate the tiles. Violations of the rules should raise the system energy so that the Penrose tiling forms a ground state. One of the failures of the quasiperiodic crystal model is that no simple, physically plausible, system is known which forces Penrose or other similar tiling rules. Indeed there is some reason to question whether such a feat is possible in principal. In an aperiodic structure each atom has an environment distinct from every other atom if far enough neighbors are considered. That suggests that the binding energy of each atom should vary. Some atoms will have low (favorable) energies and others will have slightly higher energies. If possible the system would like to minimize its energy by removing the sites of higher energy, leaving behind the low energy sites. Only in periodic structures can all sites, or a subset of sites, be absolutely equivalent in binding energy. This is the basis of arguments suggesting that ground states of matter may always be spatially periodic [14].

A further assumption is that the ground state is stable against thermal fluctuations at low temperatures. That is, there could be a phase transition out of the quasiperiodic crystal state at T = 0 into some other phase. In the projection language a two dimensional quasiperiodic crystal corresponds to a perfectly flat surface. Thermally excited phasons can cause this surface to fluctuate, but as long as the height remains bounded (in a probabilistic sense) one is still in the quasiperiodic crystal phase. If the surface roughens so that the height diverges logarithmically the system changes phase into a random tiling state. Penrose tilings with arrow rules have been studied and it appears that the roughening temperature is at T = 0 so that the Penrose tiling is unstable at all finite temperatures [15].

In three dimensions the issue is somewhat less clear because both the Penrose tiling and the random tiling have bounded fluctuations. One possible difference lies in their elasticity. The energy of a Penrose tiling (in 2-D) is known to grow linearly with phason strain, leading to a breakdown of conventional elasticity at absolute zero. If this is also true of 3-D tilings one could use the form of the elastic free energy as a distinguishing tool. If, on the other hand, the 3-D tiling does have ordinary quadratic dependence of energy on phason strain the elastic constants should be determined from the matching rule energy. In particular they must be independent of temperature. Random tiling models, on the other hand acquire their stability against phason strain from entropy. Therefore their phason elastic constants grow linearly with temperature. For the remainder of this paper I will assume that the quasiperiodic crystal model is stable at finite temperature and does possess conventional elasticity.

3. Experimental Tests

It is of great interest to test experimentally whether real equilibrium quasicrystals are better described by the random tiling model, the quasiperiodic crystal model, or perhaps neither. Several tests are conceivable. This section describes tests involving x-ray scattering, microscopy, and thermodynamic measurements. Many of the experiments have been performed already. The results generally support the random tiling model or lack sufficient resolution to be sure. What stands out most clearly is the need for further experimentation.

3.1 X-Ray diffraction

High resolution x-ray scattering probes long range order in a material. Such experiments provided the first evidence of the lack of long range order in rapidly quenched AlMn, as well as the presence of such order in AlCuFe. That is, the scattering peaks appeared to be delta functions to within instrumental resolution suggesting translational correlation lengths of more than 30000 Angstroms [16]. Diffuse scattering tails on the delta functions reveal that the ordering is not perfect despite being of infinite range.

Let us remind ourselves of the predictions of the models. The Debye-Waller effect leads to temperature dependent peak intensities

$$I = I_0 \exp\left\{\left(\frac{-T}{2}\right)\left[\frac{q_\perp^2}{K_\perp}+\frac{q_\parallel^2}{K_\parallel}\right]\right\}$$

where $K_\parallel$ and $K_\perp$ are combinations of elastic constants. The intensity lost from the peaks goes into diffuse scattering in qualitative agreement with experiments. At any finite fixed temperature the quasiperiodic crystal model and the random tiling model are indistinguishable. That is the Debye-Waller effect on peak intensities has identical form in the random tiling model and the quasiperiodic crystal model.

To distinguish the models one must vary the temperature. The quasiperiodic crystal model predicts simple exponential decay of each peak with temperature because the elastic constants are all independent of temperature. The random tiling model, on the other hand predicts that the phason elastic constants should increase linearly with temperature. This leads to instability of the quasicrystal state at low temperatures due to either coupling between phonons and phasons [17] or an instability in the phason elasticity by itself [18]. In any case the system should enter the quasicrystal phase through a first order phase transition as temperature is increased, followed by monotonic increase of intensity for peaks with large $q_{\perp}$. At high temperatures $K_{\perp}$ becomes proportional to temperature (within the harmonic approximation) so that peaks with large $q_{\perp}$ will level off at constant intensity.

Such remarkable behavior has indeed been seen in experiments by P. Bancel [16]. One note of caution is in order. Several other researchers have been unable to confirm this result, and Bancel finds it in only four out of five samples studied. Or course, to observe the effect it is essential that equilibrium be maintained throughout the experiment. Otherwise the phasons will be frozen and the large $q_{\perp}$ peaks will have constant intensity. In fact, perhaps the most remarkable finding of Bancel's experiment is the discovery that phasons can relax on laboratory time scales far below the melting temperature. Another important point is that the quasicrystal concentration regime is rather small and many attempts at confirmation have not been at precisely the correct concentration.

X-ray scattering has not yet been exploited to its fullest. In particular the diffuse scattering has not been studied. All eigenvalues of the hydrodynamic matrix can be measured, in principle, from the diffuse scattering lineshapes [19]. This information would allow unambiguous identification of an elastic instability. Perhaps it would even be possible to determine whether the instability is driven by phonon-phason coupling or is instead intrinsic to the phason degrees of freedom.

3.2 Microscopy

Microscopy attempts to image the structure and thereby give direct structural information. So far, however, these techniques have not been able to resolve the quasicrystal model issue. In particular, distinguishing between models requires observing the temperature dependence of the short wavelength phason fluctuations. But these fluctuations have not been observed at all through microscopic techniques. It is possible that they are not present in any appreciable numbers which would be a certain vindication of the quasiperiodic crystal model. But it seems far more likely that resolution and interpretation problems plague the experiments.

Lattice imaging, such as performed by Hiraga [20], show beautiful patterns with essentially no phason strain evident. But the phason strains which can be imaged clearly by this technique are long wavelength strains which show up regardless of resolution. Recall that the localized phason fluctuations do not destroy long range translational order. Therefore it is necessary to clearly resolve local atomic arrangements to observe these strains. High resolution imaging has been done, but here there is still a problem. The high resolution image is not actually a picture of atoms and is actually rather difficult to interpret.

Three effects make interpretation difficult. First of all, not all the scattered electrons are used in forming the image. Rather a ring of bright spots is recombined, excluding much of the diffuse scattering. It is well known that simply superposing plane waves with symmetrically arranged wave vectors yields pictures which greatly resemble the electron micrographs [13]. Secondly, the sample is of finite thickness so that the image is the result of averaging a structure over many atomic layers. Such averaging will tend to cancel strains of wavelength less than the sample thickness [20]. Finally, the problem of multiple scattering obscures the connection between the observed image and the real structure so that sophisticated image simulations are needed to connect the two. It would be of great interest to carry out such a calculation for a random tiling to see whether these experiments actually can observe local phason strains.

Electron microscopy is well suited to studying symmetries of materials. M. Audier [21] has performed similar analysis of AlCuFe samples annealed at 600C. His results suggest that the structure is not quasicrystalline at all, but rather microcrystalline with a unit cell of 8.9 Angstroms. This result is consistent with the observation of Bancel that the quasicrystal peaks disappear at 670C, and the observation of broadened x-ray peaks in samples annealed at 600C by Goldman [22]. These results tend to support the random tiling picture by showing that the quasicrystal is a high temperature phase while the ground state is a conventional crystal.

Scanning tunneling microscopy overcomes the problems faced by electron microscopy of connecting the atomic locations to the image. That is, the technique really does give direct information about the spatial variation of surface height with resolution adequate to locate individual atoms. So far the technique has been used [23] to demonstrate the existence of long range order, and to identify certain local atomic arrangements, but a serious search for phason fluctuations has not been made.

3.3 Thermodynamic measurements

The random tiling model predicts a few effects of a thermodynamic nature. One of these is that the system should transform from a crystal into a quasicrystal as temperature is increased. The other is that the quasicrystal phase should have an entropy significantly enhanced by tiling fluctuations. The quasiperiodic crystal model also makes a novel prediction. That is, the nonelectronic contribution to the specific heat should be enhanced by the presence of phason excitations leading to a T^3 term of twice the amplitude one would expect from phonons alone.

These predictions of the random tiling model have been confirmed by Sheild and Goldman [24]. A sample of AlCuFe was annealed at 600C then quenched to room temperature. Upon heating, a DTA scan revealed an endothermic peak at 670C. Thus the quasicrystal (high temperature) phase has a higher energy than the low temperature phase. Since the free energies must be equal at the point of phase change, the quasicrystal has a higher entropy. Of course, it is not certain that the higher entropy is the result of tiling fluctuations rather than, say, vacancies. It will be of great interest to measure the entropy change and compare to the entropies of three dimensional binary random tiling models.

4. Conclusions

This paper discusses three qualitatively different models of quasicrystalline order. The icosahedral glass model is, essentially by definition, an appropriate model for describing nonequilibrium quasicrystals. The other two models, the random tiling model and the quasiperiodic crystal model, share much in common. They both attempt to describe equilibrium structures. They both predict similar types of ordering at finite temperatures. But they are distinguishable and have a very different physical basis.

The quasiperiodic crystal model suggests that energetics favor quasiperiodicity. The ground state is a perfect quasicrystal. At finite temperature phason fluctuations are allowed, but they fail to destroy long range translational order. The random tiling model suggests that quasiperiodicity is favored by entropy. Phason fluctuations provide a source of entropy, with short wavelength phason fluctuations actually providing the mechanism for eliminating the long wavelength fluctuations and guaranteeing long range translational order.

Experimentally the models can be distinguished by x-ray diffraction, microscopy, and thermodynamic measurements. Many of these experiments have been performed. Where effects are seen the experiments tend to favor the random tiling model. But duplication of results among research groups has proven difficult and further work is needed. Experiments need to be repeated in different laboratories to ensure reproducibility. And some types of experiment, such as a study of diffuse scattering, are still needed.

I wish to acknowledge useful discussions and collaborations with C. L. Henley, M. V. Jaric, D. P. Deng, K. J. Strandburg, L. Tang, and P. Bancel. This work supported in part by the NSF under grant DMR-8613218 and by the donors of the Petroleum Research Fund.

5. References

1. D. Shechtman, I. Blech, G. Gratias, and J. W. Cahn, *Phys. Rev. Lett.* **53**, 1951 (1984)

2. D. Levine and P. J. Steinhardt, *Phys. Rev. Lett.* 53, 2477 (1984)

3. P. W. Stephens and A. I. Goldman, *Phys. Rev. Lett.* **55**, 1168 (1986)

4. C. L. Henley, *J. Phys. A* **21**, 1649 (1987)

5. M. Widom, D. P. Deng, and C. L. Henley, *Phys. Rev. Lett.* **63**, 310 (1989)

6. K. J. Strandburg, L. Tang, M. V. Jaric, *Phys. Rev. Lett.* **63**, 314 (1989)

7. V. Elser and C. L. Henley, *Phys. Rev. Lett.* **55**, 2883 (1985)

8. M. Widom, K. J. Strandburg, and R. H. Swendsen, *Phys. Rev. Lett.* **58**, 706 (1987)

9. F. Lancon and L. J. Billard, *J. Physique (Paris)* **49**, 249 (1988)

10. V. Elser in *Proceedings of the Fifteenth International Colloquium on Group Theoretical Methods in Physics*, ed. Gilmore and Feng (World Scientific, 1987)

11. T. C. Lubensky, J. E. S. Socolar, P. J. Steinhardt, P. A. Bancel, and P. A. Heiney, *Phys. Rev. Lett.* 57, 1440 (1986); P. M. Horn, W. Malzfeldt, D. P. Divincenzo, J. Toner, R. Gambino, *Phys. Rev. Lett.* 57, 1444 (1986)

12. G. Y. Onoda, P. J. Steinhardt, D. P. Divincenzo, J. E. S. Socolar, *Phys. Rev. Lett.* **60**, 2653 (1988)

13. See articles by T. C. Lubensky and by P. Bak and A. I. Goldman in *Introduction to Quasicrystals*, ed. M. V. Jaric (Academic Press, 1988)

14. P. W. Anderson, *Basic Notions of Condensed Matter Physics* (Benjamin/Cummings, 1984) pp. 12-15

15. L. Tang and M. V. Jaric, *Preprint* (1989). But note that K. J. Strandburg, Preprint (1989) finds conflicting evidence in a Monte Carlo simulation.

16. P. Bancel, *Preprint* (1989)

17. M. Widom, in *Proceedings of the Adriatico Research Conference on Quasicrystals*, ed. M. V. Jaric and S. Lundqvist (World Scientific, 1990)

18. C. L. Henley, in *Quasicrystals and Incommensurate Structures in Condensed Matter*, ed. M. Jose-Yacaman (World Scientific, 1989); Private communication (1989)

19. M. V. Jaric and D. R. Nelson, *Phys. Rev. B* **37**, 4458 (1988)

20. K. Hiraga, *These Proceedings.*

21. M. Audier, in *Proceedings of the Adriatico Research Conference on Quasicrystals*, ed. M. V. Jaric and S. Lundqvist (World Scientific, 1990)

22. A. I. Goldman, J. E. Sheild, C. A. Guryan, and P. W. Stephens, in *Proceedings of the Adriatico Research Conference on Quasicrystals*, ed. M. V. Jaric and S. Lundqvist (World Scientific, 1990)

23. A. R. Kortan, R. S. Becker, F. A. Thiel and H. S. Chen, *Phys. Rev. Lett.* **64**, 200 (1989)

24. J. E. Sheild and A. I. Goldman, *Private communication* (1989)

Phason Freezing

K. Sekimoto

Department of Physics, Kyushu University, Fukuoka 812, Japan

Abstract. We propose a macroscopic model of the freezing of phasons in quasicrystals. The phason is assumed to become frozen on the freezing front which advances in a prescribed manner. The spatial fluctuation of the frozen phason is studied and the consequences for diffraction experiments are discussed.

1. Introduction

The existence of the phason is one of the major features of quasicrystals [1]. Experimental studies have not settled the controversy about how the spatial fluctuations of the phason are suppressed, that is, whether the matching energy [2] or the configurational entropy [3] stabilizes the quasicrystals. Much effort has gone towards establishing the relation between atomic arrangements and the macroscopic, or coarse grained, small amplitude phason fluctuation [4]. The problems about quasicrystals raised from a macroscopic point of view may also be of interest as problems of more general equilibrium and non-equilibrium statistical mechanics. Both in the matching-rule model and in the random tiling model the excess free energy due to the long wavelength fluctuations of the phason should have manifest invariance [5] under the symmetry operation of quasicrystals and so does the hydrodynamic equation [6] for these fluctuations. In the present paper we consider a process of freezing of the phason using an idea which is generalizable to consider the freezing phenomena of other systems. We consider the freezing of the phason in quasicrystals through the directional cooling in which the region of (effectively) zero temperature extends into the surrounding region of high temperature in a prescribed manner (Fig.1(a)). Upon freezing the typical time scale of the dynamics of the phason becomes much larger than the time scale of the measurement. In the transient region between the frozen region and the unfrozen region the typical time scale of the phason varies both spatially and temporally. We can therefore regard the system as a collection of the coupled degrees of freedom with different time scales. Roughly speaking, those degrees of freedom with larger time scales act as the provisionally

Springer Series in Solid-State Sciences, Vol. 93 Quasicrystals
Editors: T. Fujiwara · T. Ogawa

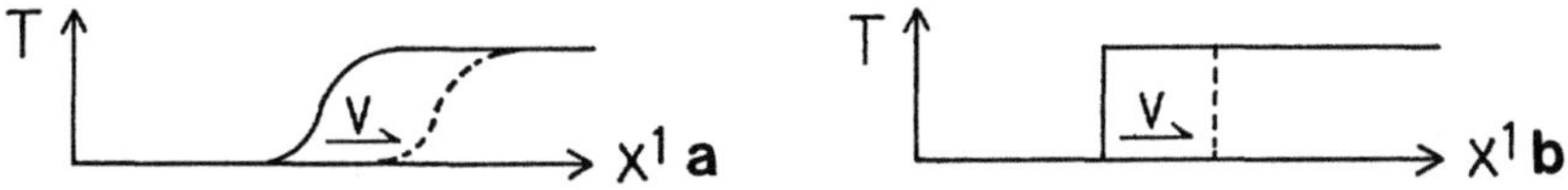

Fig. 1 (a) The temperature (T) profile that advances along the x^1-axis. (b) The same as in (a) but with a sharp freezing front. V is the front velocity.

fixed boundary conditions for the degrees of freedom with smaller time scales, while the dynamics of the former degrees of freedom is renormalized upon elimination of the latter degrees of freedom. Our interest is in the statistics of the spatial fluctuation of the phason in the frozen region.

In the present study we take up the random tiling picture of quasicrystals for which both the free energy due to phason fluctuation and the hydrodynamic equation of the phason have been well studied under several fundamental assumptions [3]. For concreteness we consider the quasicrystals with icosahedral symmetry. Furthermore, to make the calculation tractable we have assumed that (1) the frozen region and the unfrozen region are separated by a planar boundary with infinitesimal thickness, and that (2) this boundary, which we call the freezing front, is oriented in a given direction and moves with a given velocity $V > 0$ (Fig.1(b)). The directional cooling of quasicrystals may be realized when they are made by melt spinning methods or by some other *ad hoc* experiments, and the phason fluctuation can be measured by scattering techniques [7] or by high resolution electron microscopies [8]. The organization of this paper is as follows. In section 2 we describe the fluctuating hydrodynamic equation for the coarse grained phason field based on the random tiling picture. Also the boundary conditions are introduced. In section 3 we show the formal expression of the fluctuation spectra and also its expansion with respect to wave numbers. In section 4 more detailed analyses of the fluctuation spectra are done under the special choice of stiffness constants of phason elasticity. The final section (section 5) is devoted to concluding remarks.

2. Model of Phason Freezing

We ignore all hydrodynamic excitations other than phason fluctuation. Then in the random tiling picture the quadratic free energy H due to phason fluctuation can be written as follows [3]:

$$H = \frac{1}{2}\int\int\int dx^1 dx^2 dx^3 \frac{\partial w^i}{\partial x^\alpha} C_{ij}^{\alpha\beta} \frac{\partial w^j}{\partial x^\beta}. \qquad (1)$$

Here the field $(w^1, w^2, w^3) \equiv \vec{w}^T$ (the superfix T denotes transposition) is the coarse-grained phason as a function of the position $(x^1, x^2, x^3) = \boldsymbol{x}^T$ in the physical space and of the time t. We use Greek indices for the components in the physical space and Latin indices for those in the perpendicular space. We employ Einstein's summation convention for both types of indices unless otherwise stated. The stiffness tensor $C_{ij}^{\alpha\beta}$ consists of two independent contributions, say $C_{ij}^{\alpha\beta} = K_1 C_{ij}^{(1)\alpha\beta} + K_2 C_{ij}^{(2)\alpha\beta}$, where $K_1 (> 0)$ and $K_2 (> 0)$ are two independent stiffness constants [5]. The spatial fluctuation of the phason is not rough in *equilibrium* in the sense that the thermal average of $[w^i(\boldsymbol{x}) - w^i(\boldsymbol{x}')][w^j(\boldsymbol{x}) - w^j(\boldsymbol{x}')]$ remains finite in the limit of $|\boldsymbol{x} - \boldsymbol{x}'| \to \infty$ [9]. In hypothetical two-dimensional quasicrystals the corresponding quantity would diverge logarithmically with the separation $|\boldsymbol{x} - \boldsymbol{x}'|$ in equilibrium [3]. The above model predicts that the quasicrystals in three dimensions show δ function diffraction peaks [3] in equilibrium.

We assume that phason variable itself is not a conserved quantity and we adopt the simplest Ginzburg-Landau-Langevin equation for the fluctuating hydrodynamics of the phason [6]:

$$\frac{\partial}{\partial t} w^i(\boldsymbol{x}, t) = \Gamma C_{ij}^{\alpha\beta} \frac{\partial^2}{\partial x^\alpha \partial x^\beta} w^j(\boldsymbol{x}, t) + f^i(\boldsymbol{x}, t), \tag{2}$$

where Γ is the Onsager kinetic coefficient, which may depend on $\vec{w}(\boldsymbol{x}, t)$ or on the temperature field $T(\boldsymbol{x}, t)$. The last term in (2) represents thermal noise which we assume to be Gaussian white noise, whose first and second moments are, respectively,

$$< f^i(\boldsymbol{x}, t) >= 0, \tag{3}$$

$$< f^i(\boldsymbol{x}, t) f^j(\boldsymbol{x}', t') >= 2\Gamma T \delta_{ij} \delta(\boldsymbol{x} - \boldsymbol{x}') \delta(t - t'). \tag{4}$$

Hereafter the bracket $< \cdot >$ denotes the stochastic average over this noise field $f^i(\boldsymbol{x}, t)$.

We assume that the freezing front advances in the positive x^1-direction, and that the front is at the position $x^1 = Vt$ at the time t. The freezing of phason is taken into account by choosing the Onsager coefficient Γ as follows:

$$\Gamma(\boldsymbol{x}, t) = \begin{cases} 0 & \text{for} \quad x^1 \le Vt \\ \Gamma_0 & \text{for} \quad x^1 > Vt \end{cases}, \tag{5}$$

where Γ_0 is a positive constant. By this choice of Γ the boundary condition on the moving freezing front is also specified: More explicitly (5) means that $\partial w / \partial t = 0$ for $x^1 \le Vt$ and especially at $x^1 = Vt$ we can rewrite $\partial w / \partial t = 0$ as

$$V \frac{\partial w^i}{\partial x^1}\Big|_{x^1 = Vt} = \frac{d}{dt}\{w^i|_{x^1 = Vt}\}, \qquad i = 1, 2, 3, \tag{6}$$

where on the right hand side the time derivative d/dt is taken with x^2 and x^3 fixed. The above boundary condition introduces a so-called diffusion length $l_D \equiv \Gamma_0 K_0/V$ into our dynamical model, where K_0 is a typical magnitude of $C_{ij}^{\alpha\beta}$. Roughly speaking the advancing freezing front influences the phason field only within the range of the order of l_D ahead of this front. Lastly we assume that the unfrozen region ($x^1 > Vt$) is limited by the outer boundary at $x^1 = Vt + L$ with a constant $L > 0$. There we impose the phason stress-free boundary condition, i.e.

$$C_{ij}^{1\beta}\frac{\partial w^j}{\partial x^\beta}\Big|_{x^1=Vt+L} = 0, \qquad i = 1,2,3. \tag{7}$$

This choice of the boundary condition is somewhat arbitrary but this is consistent with the fact that there are no preferable values of $\vec{w}$. Beyond the outer boundary ($x^1 > Vt + L$) we can imagine either that there is no material or that the local atomic order is not established well enough to define the phason field. (In reality such an atomic disorder might induce linear phason strain through the phason-phonon coupling [5, 10].)

3. Fluctuation Spectrum of the Frozen Phason

The fluctuation spectrum of the frozen phason, $\ll \hat{w}^i(\boldsymbol{q})\hat{w}^j(\boldsymbol{q})^* \gg$, is defined as

$$(2\pi)^3\delta(\boldsymbol{q}-\boldsymbol{q}') \ll \hat{w}^i(\boldsymbol{q})\hat{w}^j(\boldsymbol{q})^* \gg \equiv < \hat{w}^i(\boldsymbol{q})\hat{w}^j(\boldsymbol{q}')^* >, \tag{8}$$

where $\hat{w}^i(\boldsymbol{q})$ is the spatial Fourier component with the wave vector $\boldsymbol{q}$ of the *frozen* phason field,

$$\hat{w}^i(\boldsymbol{q}) \equiv \int d^3\boldsymbol{x} e^{-i\boldsymbol{q}\cdot\boldsymbol{x}} w^i(\boldsymbol{x}, t = \infty), \tag{9}$$

and O^* is the complex conjugate of O. The quantity that characterizes the roughness of the frozen phason, say $< [w^i(\boldsymbol{x}) - w^i(\boldsymbol{x}')][w^j(\boldsymbol{x}) - w^j(\boldsymbol{x}')] >$, can be calculated in principle from the above spectrum. Since the value of the frozen phason at $\boldsymbol{x}^T = (x^1, x^2, x^3)$ is determined at the time $t = V^{-1}x^1$, when the freezing front passes through this point, we have to derive the evolution equation of the phason field on the moving freezing front in order to calculate $\hat{w}^i(\boldsymbol{q})$. To do so we first eliminate from the equation of motion (2) the phason field in the region $Vt < x^1 < Vt+L$ by using Green's theorem and then eliminate the phason field on the outer boundary. We finally obtain the result in the following form:

$$\begin{aligned}&\{\boldsymbol{M}_{ff}(\boldsymbol{q}) - \boldsymbol{M}_{fo}(\boldsymbol{q})[\boldsymbol{M}_{oo}(\boldsymbol{q})]^{-1}\boldsymbol{M}_{of}(\boldsymbol{q})\}\vec{\hat{w}}(\boldsymbol{q})\\ &= \vec{\Theta}_f(\boldsymbol{q}) - \boldsymbol{M}_{fo}(\boldsymbol{q})[\boldsymbol{M}_{oo}(\boldsymbol{q})]^{-1}\vec{\Theta}_o(\boldsymbol{q}),\end{aligned} \tag{10}$$

where $\boldsymbol{M}_{ff}(\boldsymbol{q})$ etc. are the 3×3 matrices which act on the vectors in the perpendicular space, and $\vec{\hat{w}}(\boldsymbol{q}) \equiv (\hat{w}^1(\boldsymbol{q}), \hat{w}^2(\boldsymbol{q}), \hat{w}^3(\boldsymbol{q}))^T$, $\vec{\Theta}_f(\boldsymbol{q}) \equiv (\Theta_f^1(\boldsymbol{q}), \Theta_f^2(\boldsymbol{q}), \Theta_f^3(\boldsymbol{q}))^T$ and similarly for $\vec{\Theta}_o(\boldsymbol{q})$. The matrices have the following integral representations:

$$\boldsymbol{M}_{BB'}(\boldsymbol{q}) = \boldsymbol{\Delta}_{BB'} + (2\pi)^{-1} \int_{-\infty}^{\infty} d\zeta \boldsymbol{F}(\boldsymbol{q}; \zeta, \Gamma_o^{-1} V) \boldsymbol{E}_{BB'}(\boldsymbol{q}; \zeta, \Gamma_o^{-1} V), \tag{11}$$

where $B = o, f$ and $B' = o, f$ and $\boldsymbol{\Delta}_{BB'} = \boldsymbol{0}$ except for $\boldsymbol{\Delta}_{ff} = \frac{1}{2}\boldsymbol{1}$. We have also defined

$$\boldsymbol{F}(\boldsymbol{q}; \zeta, \Gamma_o^{-1} V) = \{\zeta^2 \boldsymbol{C}^{11} + \zeta q_\perp \cdot \boldsymbol{C}^{\perp 1} + q_\perp q_{\perp'} \boldsymbol{C}^{\perp\perp'} + i\Gamma_o^{-1} V(q_1 - i\varepsilon - \zeta)\}^{-1}, \tag{12}$$

and $\boldsymbol{E}_{ff}(\boldsymbol{q}; \zeta, \Gamma_o^{-1} V) \equiv i(q_1 + \zeta)\boldsymbol{C}^{11} + iq_\perp(\boldsymbol{C}^{\perp 1} + \boldsymbol{C}^{1\perp}) + \Gamma_o^{-1} V \boldsymbol{1}$, $\boldsymbol{E}_{of}(\boldsymbol{q}; \zeta, \Gamma_o^{-1} V) \equiv \boldsymbol{E}_{ff}(\boldsymbol{q}; \zeta, \Gamma_o^{-1} V) e^{i\zeta L}$, $\boldsymbol{E}_{oo}(\boldsymbol{q}; \zeta, \Gamma_o^{-1} V) \equiv i\zeta \boldsymbol{C}^{11} + iq_\perp \boldsymbol{C}^{\perp 1} + \Gamma_o^{-1} V \boldsymbol{1}$, and $\boldsymbol{E}_{fo}(\boldsymbol{q}; \zeta, \Gamma_o^{-1} V) \equiv \boldsymbol{E}_{oo}(\boldsymbol{q}; \zeta, \Gamma_o^{-1} V) e^{-i\zeta L}$, where $\varepsilon = 0+$ and $(\boldsymbol{C}^{11})_{ij} \equiv C_{ij}^{11}$ etc. and the repeated indices $\perp$ and $\perp'$ mean to take the summation over the components in the physical space except x^1-component. $\boldsymbol{1}$ is the 3×3 unit matrix. $\vec{\Theta}_f(\boldsymbol{q})$ and $\vec{\Theta}_o(\boldsymbol{q})$, are the linear functional of the noise field and are given as follows:

$$\vec{\Theta}_f(\boldsymbol{q}) = \int_{-\infty}^{\infty} dt_o \int_0^L dl_o \int_{-\infty}^{\infty} \frac{d\zeta}{2\pi} \int d^2 \boldsymbol{x}^\perp e^{-i[\zeta l_o + (q_1 - i\varepsilon)V t_o + q_\perp \cdot x^\perp]} \Gamma_o^{-1} V \boldsymbol{F}(\boldsymbol{q}; \zeta, \Gamma_o^{-1} V) \vec{f}(x^1 = l_o + V t_o, x^2, x^3, t_o). \tag{13a}$$

$$\vec{\Theta}_o(\boldsymbol{q}) = \int_{-\infty}^{\infty} dt_o \int_0^L dl_o \int_{-\infty}^{\infty} \frac{d\zeta}{2\pi} \int d^2 \boldsymbol{x}^\perp e^{-i[\zeta(L - l_o) + (q_1 - i\varepsilon)V t_o + q_\perp \cdot x^\perp]} \Gamma_o^{-1} V \boldsymbol{F}(\boldsymbol{q}; \zeta, \Gamma_o^{-1} V) \vec{f}(x^1 = l_o + V t_o, x^2, x^3, t_o). \tag{13b}$$

The equation (10) is just the evolution equation for the boundary values of the phason field on the freezing front in the Fourier transformed form. The correspondences to the real form are $iq_1 \leftrightarrow V\partial/\partial t$ and $i\boldsymbol{q}_\perp \equiv (iq_1, iq_2) \leftrightarrow (\partial/\partial x^1, \partial/\partial x^2)$. This equation describes the non-Markov process since the eliminated part of the phason field yields memory terms. From eqs.(4) and (10) we can obtain the closed expression for the fluctuation spectrum (8). But in general the integrals in (11) are hard to evaluate, and then we study their behavior in the limit of small $|\boldsymbol{q}|$, which is relevant to the behavior of $< [w^i(\boldsymbol{x}) - w^i(\boldsymbol{x}')][w^j(\boldsymbol{x}) - w^j(\boldsymbol{x}')] >$ in the limit of $|\boldsymbol{x} - \boldsymbol{x}'| \to \infty$. We shall show only the resulting approximate expression for (10) to the lowest order of $|\boldsymbol{q}|$,

$$\{i\Gamma_o C^{1\beta} q_\beta + O(q^2)\}\vec{w}(\boldsymbol{q}) = \vec{\Theta}_o(\boldsymbol{q}=0) + O(q). \tag{14}$$

From (4),(13b) and (14) the fluctuation spectra can be shown to be $\ll \vec{w}(\boldsymbol{q})\vec{w}(\boldsymbol{q})^\dagger \gg = T[q_\alpha C^{\alpha 1}(C^{11}[1-\exp\{-2L\Gamma_o^{-1}V(C^{11})^{-1}\}])^{-1}C^{1\beta}q_\beta + O(q^4)]^{-1}$ where $(\vec{A}\vec{B}^\dagger)^{ij} \equiv A^i B^{j*}$. Here we convert our coordinate system so far used into one which is not dependent on the freezing direction. For this we have only to employ the following abstract notations $< \boldsymbol{a}|\boldsymbol{C}|\boldsymbol{b} >$ for $a_\alpha C^{\alpha\beta} b_\beta$, for example $q_\alpha C^{\alpha 1} \rightarrow < \boldsymbol{q}|\boldsymbol{C}|\hat{\boldsymbol{n}} >$, where $\hat{\boldsymbol{n}}$ is the unit vector along the freezing direction. Such replacements yield the formula for the fluctuation spectra of the frozen phason in the long wavelength limit as follows:

$$\begin{aligned} &\ll\vec{w}(\boldsymbol{q})\vec{w}(\boldsymbol{q})^\dagger \gg = T[< \boldsymbol{q}|\boldsymbol{C}|\hat{\boldsymbol{n}} > (< \hat{\boldsymbol{n}}|\boldsymbol{C}|\hat{\boldsymbol{n}} > \\ &\quad [1-\exp\{-2L\Gamma_o^{-1}V < \hat{\boldsymbol{n}}|\boldsymbol{C}|\hat{\boldsymbol{n}} >^{-1}\}])^{-1} < \hat{\boldsymbol{n}}|\boldsymbol{C}|\boldsymbol{q} > + O(q^4)]^{-1}. \end{aligned} \tag{15}$$

The lowest order approximation in the long wavelength limit would be to ignore the part written as $O(q^4)$ on the right hand side of (15). If we do so, the spectrum becomes $\propto |\boldsymbol{q}|^{-2}$ just like the equilibrium spectrum [9] and here the spectrum depends on V through the ratio L/l_D. But in fact in this approximation the spectrum *diverges* for some directions of $\boldsymbol{q}$ along

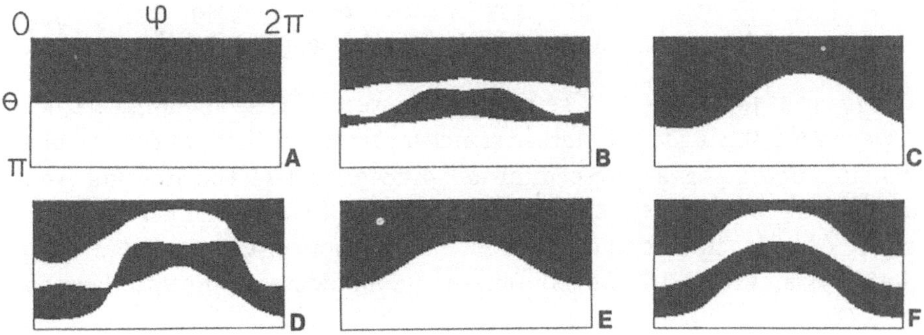

Fig. 2
The regions where the determinant $\det(< \boldsymbol{q}|\boldsymbol{C}|\hat{\boldsymbol{n}} >)$ is positive (black) or negative (white) are shown on the plane representing the directions of $\boldsymbol{q}$ for several choices of $\hat{\boldsymbol{n}}$ and also of K_1/K_2 (see the text). The directions of $\boldsymbol{q}$ and $\hat{\boldsymbol{n}}$ are specified by the polar representation, in which one of the 5-fold axes of the quasicrystal is along the $\theta = 0$ axis and another 5-fold axis is on the $\phi = 0$ plane. The coordinate systems of all these figures are common and shown only in Fig.2a. In Figs. 2a,c,e we have chosen $K_1/K_2 = 1$ while in the Figs. 2b, d, f we have chosen $K_1/K_2 = 3$. The freezing direction $\hat{\boldsymbol{n}}$ is chosen along a 5-fold axis ($\theta = 0$) in Figs.2a, b, along a 3-fold axis ($\theta = \sin^{-1}(2/\sqrt{3\tau+6}), \phi = \pi/5$), in Figs.2c, d, and along a 2-fold axis ($\theta = \sin^{-1}(1/\sqrt{\tau+2}), \phi = 0$) in Figs.2e, f, where $\tau \equiv (1+\sqrt{5})/2$.

which the matrix $< \boldsymbol{q}|C|\hat{\boldsymbol{n}} >$ becomes secular. This means that along such directions the spectrum is not $O(|\boldsymbol{q}|^{-2})$ but is singularly enhanced to become at least $O(|\boldsymbol{q}|^{-4})$. We show in Fig.2 these singular directions of $\boldsymbol{q}$ as the boundaries of the black and white regions for three choices of $\hat{\boldsymbol{n}}$'s directed along the high symmetry directions of the icosahedral quasicrystals.

4. Phason Roughness for a Special Choice of the Stiffness Constants

We analyzed in detail the specially symmetric (isotropic) case of $K_1 = K_2$, that is $\boldsymbol{C}^{\alpha\beta} = K_o \delta_{\alpha\beta} \mathbf{1}$. In this case all the components of the frozen phason, $\hat{w}^i(\boldsymbol{q})$, become mutually independent (see eqs. (1)-(7)), and we can obtain the explicit expression for $\ll |\hat{w}^1(\boldsymbol{q})|^2 \gg = \ll |\hat{w}^2(\boldsymbol{q})|^2 \gg = \ll |\hat{w}^3(\boldsymbol{q})|^2 \gg$. In the limit of either $l_o \ll L \ll |\boldsymbol{q}|^{-1}$ or $L \ll l_D \ll |\boldsymbol{q}|^{-1}$ the result becomes especially simple and we have the following expressions:

$$\ll |w^1(\boldsymbol{q})|^2 \gg \approx \frac{T}{K_o} \frac{N}{(q_1)^2 + \lambda^2(|\boldsymbol{q}_\perp|^2)^2}, \tag{16}$$

where

$$\begin{cases} N = 1 & \text{and} \quad \lambda = l_D \quad \text{for} \quad l_D \ll L \ll |\boldsymbol{q}|^{-1} \\ N = l_o^{-1} L & \text{and} \quad \lambda = L \quad \text{for} \quad L \ll l_D \ll |\boldsymbol{q}|^{-1}, \end{cases} \tag{17}$$

and $|\boldsymbol{q}_\perp|^2 \equiv |\boldsymbol{q}|^2 - (q_1)^2$. The asymptotic form (16) seems quite natural if we regard this as the *dynamic* structure factor of the two-dimensional system undergoing a phase diffusion process, where the freezing front memorizes continuously the unfrozen phason field in the region $Vt < x^1 < Vt + L$, and behind the front the memories are expanded along the x^1-axis. From (16) we can obtain the following asymptotic laws for $|x| \to \infty$,

$$< |w^1(\boldsymbol{x}) - w^1(\boldsymbol{o})|^2 > \approx$$
$$\begin{cases} \frac{T}{K_o} \frac{1}{2\pi l_D} & \log \frac{\sqrt{\lambda |x^1|}}{a_o} & \text{for } \lambda |x^1| \gg |\boldsymbol{x}^\perp|^2 \\ \frac{T}{K_o} \frac{1}{2\pi l_D} & \log \frac{|x^\perp|}{a_o} & \text{for } \lambda |x^1| \ll |\boldsymbol{x}^\perp|^2. \end{cases} \tag{18}$$

Here again $\lambda = \min(l_D, L)$ and we have defined $|\boldsymbol{x}^\perp|^2 \equiv |\boldsymbol{x}|^2 - (x^1)^2$ and a_o is a microscopic cut-off length. These logarithmic divergences lead to the power law singularity of the diffraction peaks in place of the δ function peaks obtained in the case of the equilibrated phason [11]. In the more general case of $K_1 \neq K_2$ we expect that the quantity $< [w^i(\boldsymbol{x}) - w^i(\boldsymbol{o})][w^j(\boldsymbol{x}) - w^j(\boldsymbol{o})] >$ will be at least logarithmically divergent since for $K_1 \neq K_2$ the spectrum $\ll \vec{w}(\boldsymbol{q})\vec{w}^\dagger(\boldsymbol{q}) \gg$ becomes $O(q^{-4})$ for more directions of $\boldsymbol{q}$ as compared with the case of $K_1 = K_2$, as seen from Fig.1. If this is the case, we have either power-law shaped peaks or possibly diffusive peaks of diffraction due to the phason fluctuation for a general choice of K_1 and K_2. Returning to the special isotropic case, we have done a similar calculation for one- and two-dimensional models. In one dimension the quantity $< |w^1(x) - w^1(o)|^2 >$ grows linearly with the separation $|x|$, and this is just as expected from the classical work by Hendricks and Teller [12]. In two dimensions we assumed that the fluctuation spectrum is given by (16), where $q_\perp = q_2$, and we have calculated the roughness exponent ζ defined by $< |w^1(\boldsymbol{x}) - w^1(\boldsymbol{o})|^2 > \sim |\boldsymbol{x}|^{2\zeta}$ for $|\boldsymbol{x}| \to \infty$. (Note that in this definition $\zeta = 1/2$ in the one-dimensional model.) We have obtained anisotropic exponents for the two-dimensional model; $\zeta = 1/4$ for $\lambda|x^1| \gg |x^2|^2$ and $\zeta = 1/2$ for $\lambda|x^1| \ll |x^2|^2$. This means that the spatial fluctuation of phason is singularly suppressed in the direction along which the freezing front advances. The prefactor of $|\boldsymbol{x}|^{2\zeta}$ in the above definition is roughly proportional to V for both $l_D \ll L$ and $l_D \gg L$ in either two or three dimensions, and in one dimension the prefactor is $(T/K_o)(1 - e^{-2L/l_D})$. Then for any of these dimensionalities the frozen fluctuation decreases if we reduce the front velocity V. This result may seem qualitatively natural since we usually expect that the slower cooling process suppresses more of the fluctuationsthat existed when the system was at high temperature. But as we assumed that there is a sharp freezing front, the use of the word 'slow' needs some caution. More microscopic analysis will clarify the relation between the macroscopic dynamics and the microscopic kinetics of freezing on the freezing front.

5. Concluding Remarks

We have analyzed the frozen fluctuation of the phason in a macroscopic model under some simplifying assumptions. In real quasicrystals the phason may or may not be completely frozen depending both on temperature and on inter-atomic potentials [13]. In cases where there are transitions to other structures [13], our analysis will need to be extended to the non-linear (phason) elasticity [14]. The nonlinear elasticity in solids sometimes leads to quite fascinating phenomena [15]. Thus the *in situ* observation of phason dynamics during heat treatment should be strongly encouraged.

We have not considered how the axes of symmetry of quasicrystals are oriented depending on the freezing direction nor have we considered the generation of linear phason strain [10], both of which should be taken into account to improve the present work. It will also be interesting to study the directional freezing of the order parameters in systems other than quasicrystals using the notion of the freezing boundary condition.

Acknowledgements

The author has benefited from valuable discussions with the participants of this Symposium, and especially thanks C. Henley and M. Jarić for their critical comments. He also acknowledges T. Kawakatsu and Y. Ishii for their helpful comments. This work is supported by the Scientific Research Fund of the Ministry of Education, Science and Culture. Finally he is grateful to the Taniguchi Foundation for presiding over this Symposium.

References

1. P. A. Kalugin, A. Y. Kitaev, and L. S. Levitov, Pis'ma Zh. Eksp. Teor. Fiz. **41**, 119 (1985).
2. G. Y. Onoda, P. J. Steinhardt, D. P. DiVincenzo, and J. E. Socolar, Phys. Rev. Lett. **60**, 2653 (1988);J. E. Socolar, these Proceedings; P. J. Steinhardt, these Proceedings.
3. C. L. Henley, J. Phys. A**21**, 1649 (1988); M. Widom, D. P. Deng, and C. L. Henley, Phys. Rev. Lett. **63**, 310 (1989); K. J. Strandburg, L.-H. Tang and M. V. Jarić, Phys. Rev. Lett. **63**, 314 (1989); M. Widom, these Proceedings.
4. M. Peryard and S. Aubry, J. Phys. C**16**, 1593 (1983); D. M. Frenkel, C. L. Henley, and E. D. Siggia, Phys. Rev. B**34**, 3649 (1986); L. S. Levitov, J. Physique **50**, 3181 (1989).
5. D. Levine, T. C. Lubensky, S. Ostlund, S. Ramaswamy, and P. J. Steinhardt, Phys. Rev. Lett. **54**, 1520 (1985); P. Bak, Phys. Rev. B**32**, 5764 (1985); P. A. Kalugin, A. Y. Kitaev, and L. S. Levitov: J. Physique Lett. **46**, L601 (1985).
6. T. C. Lubensky, S. Ramaswamy, and J. Toner, Phys. Rev. B**32**., 7444 (1985).
7. T. C. Lubensky, J. E. Socolar, P. J. Steinhardt, P. A. Bancel and P. A. Heiney, Phys. Rev. Lett. **57**, 1440 (1986).
8. K. Hiraga, M. Hirabayashi, A. Inoue, and T. Matsumoto, Sci. Rep. Res. Inst. Tohoku Univ. A**32**, 309 (1985); K. Hiraga, these Proceedings.
9. C. L. Henley, Comments Cond. Matter Phys. **13**, 59 (1987).
10. L. V. Mikheev, Phys. Lett. A**132**, 137 (1988).
11. A. Caillé, C. R. Acad. Sci. Ser. B**274**, 891 (1972).
12. S. Hendricks and E. Teller, J. Chem. Phys. **10**, 147 (1942).
13. M. V. Jarić and U. Mohanty, Phys. Rev. Lett. **58**, 230 (1987); M. Widom, these Proceedings; C. L. Henley, preprint.
14. Y. Ishii, these Proceedings.
15. K. Sekimoto, N. Suematsu, and K. Kawasaki, Phys. Rev. A**39**, 4912 (1989).

Elastic Instability Induced by Phasons in Icosahedral Quasicrystals

Y. Ishii

Institute for Solid State Physics, University of Tokyo,
Minato-ku, Tokyo 106, Japan

Abstract. Structural modulation induced by phason degrees of freedom in icosahedral quasicrystals is studied theoretically. The non-vanishing phason strains break the global symmetry of a system into that of a subgroup of the icosahedral point group. All the possible ways of the structural modulation are derived. In order to examine the possibility of such a symmetry change in icosahedral quasicrystals, the elastic free energy involving higher order terms in the phason strains is investigated. It is found that cooperative freezing of the Γ_{4g} and Γ_{5g} phason modes deforms the cubic and pentagonal modulated phases into lower symmetry phases while the rhombohedral phase might be stable. The possible phase diagrams are discussed.

1. Introduction

Quasicrystal is an incommensurate solid with a crystallographically forbidden rotational symmetry, under which we have irrational length scales along one or more directions making the translational order *quasiperiodic* rather than *periodic* [1]. In incommensurate systems, dynamics at low temperatures is described by the hydrodynamical (Goldstone) modes due to the incommensurability (*phasons*) as well as those due to the broken translational symmetry (*phonons*) [2]. In the density wave picture of quasicrystals, analogous description of low energy dynamics in terms of phonons and phasons can be made although the unit cell (tiling) picture suggests controversial phason dynamics [3,4]. In this article we adopt the former point of view.

Icosahedral quasicrystalline structure is described as an ordered phase of six density wave order parameters $\rho_{\mathbf{G}}$ where the reciprocal lattice vector $\mathbf{G}$ is in one of the six vertex directions of an icosahedron [5-7]. Although more density waves are needed for accurate descriptions of energetics based on Landau theory of crystalization [7,8], essential features of low energy dynamics can be obtained by the six order parameters. Since the mass density of atoms given by

$$\rho(\mathbf{r}) = \sum_{\mathbf{G}} \rho_{\mathbf{G}} \exp(i\mathbf{G}\cdot\mathbf{r}) = \sum_{\mathbf{G}} |\rho_{\mathbf{G}}| \exp(i\phi_{\mathbf{G}}) \exp(i\mathbf{G}\cdot\mathbf{r}) \tag{1}$$

should be real, $\rho_{\mathbf{G}}$ satisfies the relation $\rho^*_{\mathbf{G}} = \rho_{-\mathbf{G}}$, or $\phi_{\mathbf{G}} = -\phi_{-\mathbf{G}}$.

Usually amplitude fluctuations of the order parameters are suppressed at low temperatures. Then the important degrees of freedom will be the phase fluctuations of the six density wave order parameters. The six independent phases of the order parameters are expressed alternatively in terms of the six displacement fields as [5]

$$\phi_{\mathbf{G}} = \sum_{i=1}^{3} G_{\parallel}^{(i)} u_i + \sum_{i=1}^{3} G_{\perp}^{(i)} v_i, \tag{2}$$

where $\mathbf{G} = (G_{\parallel}^{(1)}, G_{\parallel}^{(2)}, G_{\parallel}^{(3)})$ and $\mathbf{G}^{\perp} = (G_{\perp}^{(1)}, G_{\perp}^{(2)}, G_{\perp}^{(3)})$ is its conjugate vector in an imaginary (perpendicular) space. u_i is the displacement in the real (physical) space relevant to the phonon degrees of freedom and v_i is the displacement in the perpendicular space relevant to the phason degrees of freedom.

The spatially uniform displacements contribute to a change in phase of the density waves. So the spatially uniform displacements do not modify the diffraction patterns, which are determined by $|\rho_{\mathbf{G}}|$. Uniform displacements in the physical space induce the rigid translation whereas those in the perpendicular space cause a transformation into a *locally isomorphic* quasicrystal [1,3]. If we assume that the free energy does not depend on the phase degrees of freedom of the order parameters, locally isomorphic quasicrystals are energetically degenerate. Then the translational symmetry along the perpendicular space is broken in the quasicrystalline ground state and phasons are Goldstone modes associated with this symmetry breaking.

Spatially non-uniform displacement fields, on the other hand, cost structural energy and their energetics is given by generalized elasticity theory including both the phonon and phason degrees of freedom [5]. The non-uniform displacement in the physical space gives the ordinary elastic deformation while that in the perpendicular space modifies the ideal arrangement of structural units locally [3]. If the characteristic wavelength of this phase fluctuation is as long as a sample size, the system should be regarded as modulated quasicrystals with a uniform phason strain. It is known that a registered commensurate phase usually is observed in a conventional incommensurate system with competing interactions [2]. So a modulated quasicrystal with the frozen long wavelength phase fluctuations, which is a registered phase of quasicrystals, is plausible in real materials. Actually, complicated crystalline phases of alloys are found in a similar stoichiometry to the icosahedral phases [9-12].

In this paper, we shall discuss the possible structural modulation caused by the non-uniform phason displacement fields and its energetic stability. A part of the discussions in this article has been published in refs.[12,13].

2. Structural Modulations in Icosahedral Quasicrystals

Symmetry of the icosahedral quasilattice is given by the 120-element icosahedral point group, $I_h = I \times C_i$ (Table 1). Here vectors in the real and perpendicular spaces are transformed according to the irreducible representations of I_h, Γ_{3u} and Γ'_{3u}, respectively [6,7]. So the spatial derivatives of the phonon displacements are transformed as

$$\Gamma_{3u} \times \Gamma_{3u} = \Gamma_{1g} + \Gamma_{3g} + \Gamma_{5g}, \tag{3a}$$

where the first, second and third terms represent dilatation, rotation and shear strains, respectively. The spatial derivatives of the phason displacements are transformed as

$$\Gamma_{3u} \times \Gamma'_{3u} = \Gamma_{4g} + \Gamma_{5g}. \tag{3b}$$

Table 1 : Character table of icosahedral point group I.

	E	$12C_5$	$12C_5^2$	$20C_3$	$15C_2$
Γ_1	1	1	1	1	1
Γ_3	3	τ	$-\tau^{-1}$	0	–1
Γ'_3	3	$-\tau^{-1}$	τ	0	–1
Γ_4	4	–1	–1	1	0
Γ_5	5	0	0	–1	1

By our definition of the coordinate axes, the normal modes for the phason strains are given as

$$
\begin{aligned}
f_1 &= (\partial_1 v_1 + \partial_2 v_2 + \partial_3 v_3)/\sqrt{3}, \\
f_2 &= (\tau^{-1}\partial_1 v_2 + \tau\partial_2 v_1)/\sqrt{3}, \\
f_3 &= (\tau^{-1}\partial_2 v_3 + \tau\partial_3 v_2)/\sqrt{3}, \\
f_4 &= (\tau^{-1}\partial_3 v_1 + \tau\partial_1 v_3)/\sqrt{3},
\end{aligned}
\tag{4}
$$

for the Γ_{4g} mode and

$$
\begin{aligned}
g_1 &= (-\tau\partial_1 v_1 + \tau^{-1}\partial_2 v_2 + \partial_3 v_3)/2, \\
g_2 &= (-\tau^{-2}\partial_1 v_1 + \tau^2\partial_2 v_2 - \sqrt{5}\partial_3 v_3)/2\sqrt{3}, \\
g_3 &= (\tau\partial_1 v_2 - \tau^{-1}\partial_2 v_1)/\sqrt{3}, \\
g_4 &= (\tau\partial_2 v_3 - \tau^{-1}\partial_3 v_2)/\sqrt{3}, \\
g_5 &= (\tau\partial_3 v_1 - \tau^{-1}\partial_1 v_3)/\sqrt{3},
\end{aligned}
\tag{5}
$$

for the Γ_{5g} mode. Here τ is a golden mean.

The phason strains are transformed according to the higher dimensional irreducible representations. This implies that, with the non-vanishing phason strains, the system is not invariant under all the symmetry operations of I_h any more. Then the icosahedral symmetry of the system changes into the symmetry of its subgroup. Since the phason strain does not violate the inversion symmetry, possible symmetry changes are $I_h \to T_h, D_{5d}, D_{3d}, D_{2h}, S_6$ and C_{2h}. It should be stressed here that the phason fields do not introduce any additional symmetry. In this sense the present analysis is valid in the vicinity of the icosahedral phase, i.e., valid for weakly first order or higher order phase transitions caused by the phason degrees of freedom.

We can determine which modes are responsible for a particular symmetry change with the help of group theory. To illustrate how to do this, we discuss the symmetry change of $I_h \to D_{3d}$. By reducing the Γ_{4g} and Γ_{5g} representations to the irreducible representations of D_{3d}, we find that both Γ_{4g} and Γ_{5g} contain one identity representation of D_{3d}. This means that one component of the Γ_{4g} mode and one of the Γ_{5g} can be made invariant under the symmetry operations of D_{3d}. These components of the phason strains do not necessarily vanish in the D_{3d} symmetry phase. Considering that the symmetry of the system under any non-vanishing phason strains is not the icosahedral one any more, we conclude that the invariant components cause a symmetry change $I_h \to D_{3d}$. The invariant components can be obtained explicitly by using the projection operator technique [14]. The results for all the possible symmetry changes are shown in Table 2. Note that we have other equivalent choices of the non-vanishing phason strains because there are several ways of changing symmetry. For example, we have ten equivalent expressions for the phason strains inducing the $I_h \to D_{3d}$ symmetry change corresponding to the choice of the three-fold axis.

The non-vanishing phason strain contributes to shifting Bragg spots as

$$
G_{\parallel}^{(i)\prime} = G_{\parallel}^{(i)} + \sum_{j=1}^{3}(\partial_i v_j)G_{\perp}^{(j)}. \tag{6}
$$

If we have complete knowledge about the mass density waves $\rho_{\mathbf{G}}$ up to the higher harmonics, detailed real space structure of the strained quasicrystal is obtained by the Fourier transformation as

$$
\rho'(\mathbf{r}) = \sum_{\mathbf{G}} \rho_{\mathbf{G}} \exp(i\mathbf{G}' \cdot \mathbf{r}). \tag{7}
$$

Table 2 : Possible modulated icosahedral phases.

Symm.	Mode	Invariant Component	$\mathcal{I}_4$	$\mathcal{I}_5$	$\mathcal{I}'_5$
T_h	Γ_{4g}	$(1,0,0,0)$	$\sqrt{3}/2$	0	0
D_{5d}	Γ_{5g}	$(\sqrt{5},\sqrt{3},2\sqrt{3},0,0)$	0	$2/\sqrt{5}$	0
D_{3d}	Γ_{4g} Γ_{5g}	$(1,\sqrt{5},0,0)$ $(\sqrt{3},-\sqrt{5},2,0,0)$	$1/3\sqrt{2}$	0	2/3
D_{2h}	Γ_{4g} $2\Gamma_{5g}$	$(1,0,0,0)$ $(1,0,0,0,0)$ $(0,1,0,0,0)$	— †	$\sqrt{2}/4$	$\sqrt{6}/4$
S_6	$2\Gamma_{4g}$ Γ_{5g}	$(1,\sqrt{5},0,0)$ $(0,0,1,-1)$ $(\sqrt{3},-\sqrt{5},2,0,0)$	no	— †	— †
C_{2h}	$2\Gamma_{4g}$ $3\Gamma_{5g}$	$(1,0,0,0)$ $(0,1,0,0)$ $(1,0,0,0,0)$ $(0,1,0,0,0)$ $(0,0,1,0,0)$	no	no	no

† Same invariant components as in the higher symmetric cases.

Even if the information is restricted to only a few of the density waves, we are still able to calculate a high resolution lattice image as in the electron microscope technique [3].

However, it is possible here to generate the real space structure more directly. We consider the three dimensional (3d) Penrose lattice, which is a geometrical model for the icosahedral quasicrystal. The strained 3d Penrose lattice is obtained by a modified projection scheme, in which the acceptance region or the cutting plane is rotated uniformly in the six dimensional (6d) space. More precisely let us define the 6d orthonormal vector set $\{|p_i\rangle, i = 1,6\}$. Here the real space is spanned by $\{|p_i\rangle, i = 1,3\}$ while $\{|p_i\rangle, i = 4,6\}$ spans the perpendicular one. The modulated quasicrystal structure is generated by using the modified base vectors $\{|p'_i\rangle, i = 4,6\}$ spanning the perpendicular space given by

$$\begin{pmatrix} |p'_4\rangle \\ |p'_5\rangle \\ |p'_6\rangle \end{pmatrix} = \underline{C}_\perp \begin{pmatrix} |p_4\rangle \\ |p_5\rangle \\ |p_6\rangle \end{pmatrix} - \underline{S} \begin{pmatrix} |p_1\rangle \\ |p_2\rangle \\ |p_3\rangle \end{pmatrix} \tag{8}$$

where

$$\underline{C}_\perp = \sum_{n=0}^{\infty} \frac{(-1)^n}{(2n)!} (\underline{\Phi}^t \underline{\Phi})^n, \qquad \underline{S} = \sum_{n=0}^{\infty} \frac{(-1)^n}{(2n+1)!} (\underline{\Phi}^t \underline{\Phi})^n \underline{\Phi}^t. \tag{9}$$

Here $\underline{\Phi} = (\phi_{ij})$ is a 3×3 matrix of rotation angles given by $\phi_{ij} = \partial_i v_j$ and $\underline{\Phi}^t$ is a transpose of $\underline{\Phi}$. If we assume that the base vectors spanning the physical space are not modified, the resulting modulated structure consists of the ideal rhombohedral tiling units.

$\underline{C}_\perp$ and $\underline{S}$ can be calculated explicitly by using the non-vanishing phason strains given in Table 2. For the case of the D_{2h} phase, for example, we have $\partial_i v_j = \delta_{ij}\phi_i$ because f_1, g_1 and g_2 do not vanish. Then $\underline{C}_\perp$ and $\underline{S}$ are given as

$$\left(\underline{C}_\perp\right)_{ij} = \delta_{ij} \cos\phi_i, \qquad \left(\underline{S}\right)_{ij} = \delta_{ij} \sin\phi_i. \tag{10}$$

The results for the other modulated phases with the T_h, D_{5d}, and D_{3d} symmetries are given in ref.[13].

The number of the non-vanishing phason strains in the modulated phase is related to the number of incommensurate ratios inherent in icosahedral geometry. For example, three independent phason strains in the D_{2h} symmetry phase, ϕ_i, are related to incommensurate ratios along three two-fold axes orthogonal to each other in an icosahedron. If the amplitude of the phason strain is locked to a special value giving the rational ratios, the modulated phase has a periodic structure regarded as a *registered* commensurate phase. In some low-dimensional incommensurate systems, the registered phases with different periodicities emerge successively [2]. Although the successive transition between several registered phases would not be so likely in the present system, it is reasonable to expect to find the modulated structure near the icosahedral phase in the phase diagram. Actually several complicated crystalline phases in a similar stoichiometry to the icosahedral one have been reported [9,10].

When the non-vanishing amplitude of phason strains in the D_{2h} phase is locked to a special value as

$$\phi_i = \arctan \frac{p_i - q_i\tau}{p_i\tau + q_i}, \tag{11}$$

where p_i/q_i is a rational approximation to the irrational ratio τ, the modulated structure is a periodic one with an orthorhombic unit cell. By choosing $p_i/q_i = 1/1, 2/1$ and $3/2$ and assuming that the edge length of the rhombohedral tiles equals to 0.45 (nm), we obtain the lattice constants as $a = 1.24, b = 2.00$ and $c = 3.24$ (nm), which are similar to the experimental values for orthorhombic Al-Cr alloy [15]. The real space structure for this D_{2h} phase is shown in Fig.1. Although we are not sure whether the orthorhombic phase of Al-Cr alloy can be interpreted as this modulated Penrose lattice because the atomic positions for the Al-Cr alloy are unknown, the coincidence of the lattice parameters is very interesting.

Another example of the structural modulation is a specific configuration of the quenched phason strains explaining asymmetric shapes of Bragg peaks. Socolar and Wright [16] assumed a mosaic structure of strained quasicrystals to explain the asymmetric shapes of Bragg spots. The phason fields introduced for a single pyramidal domain spanned by triadic vertex vectors of an icosahedron give the D_{3d} modulated phase. Although we have two independent phason strains in the D_{3d} phase, the phason fields assumed by Socolar and Wright are described by a single parameter with a special amplitude ratio of the Γ_{4g} and Γ_{5g} modes. This special choice is realized also in the rational approximation, which gives a commensurate phase with a prolate rhom-

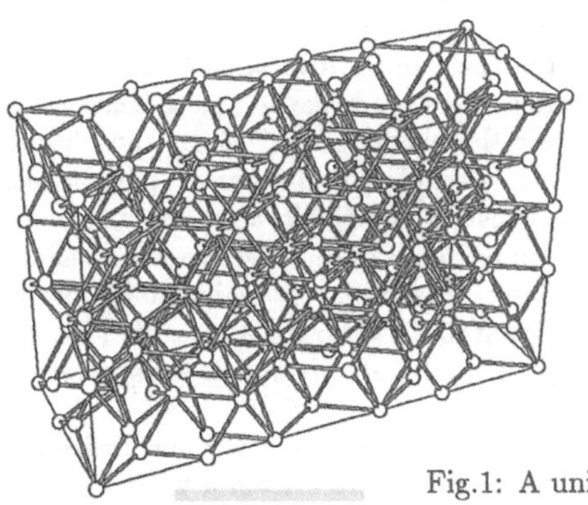

Fig.1: A unit cell structure of the D_{2h} modulated phase.

bohedral unit cell. Therefore we think that the instability against the rhombohedral deformation is inherent in the quasicrystals observed here and that a transition to some commensurate phases is physically plausible.

The phason fields do not introduce any additional symmetry. However, we know several exotic crystalline alloy phases with hexagonal symmetry (e.g. α(AlFeSi) [11,17] and μ(AlMn) [18]), where the non-icosahedral mirror symmetry is found perpendicular to the three-fold axis. Then analysis of these hexagonal systems in terms of the D_{3d} modulation is not straightforward and we should introduce a stacking fault or other defects. Decagonal phases of Al-TM (transition metal) alloys, which are considered as closely related to the icosahedral phases, also involves the non-icosahedral mirror symmetry [19]. In fact the D_{5d} modulation caused by the non-vanishing Γ_{5g} phason strain can generate a periodic structure along a five-fold axis. In this case, the mirror symmetry perpendicular to the five-fold axis can be introduced by elongating the rhombohedral units along the five-fold axis with a special choice of the Γ_{5g} phason fields [12].

3. Energetic Stability of Modulated Phases

The phason is an elementary excitation from the quasicrystalline ground state. Its energetics is obtained if the amplitude fluctuation of the icosahedral density wave order parameters can be neglected at low temperatures. The structural free energy is then given in terms of the phonon and phason displacements. By definition, the free energy does not depend on the displacement fields explicitly. So the quadratic terms of the phonon and phason strains, which give the linear elasticity, are the lowest relevant terms.

The minimal couplings in the free energy can be derived by using the group theoretical method. For the n-th order terms, we should find all the invariant products of n components of the elastic and phason strains under the icosahedral symmetry operations. The number of such invariants is given by the number of times which the identity representation is contained in a symmetric part of the product representation [20]. The minimal couplings are obtained explicitly again by using the projection operator technique although the calculation is surely tedious for higher order terms.

The quadratic (bilinear) terms give the generalized elasticity theory including the phason degrees of freedom [5]. We have two independent elastic constants, which are given conventionally in terms of the Lamè coefficients. So the mechanical properties of the icosahedral quasicrystals are isotropic. For the phason degrees of freedom, we have two stiffness coefficients for the Γ_{4g} and Γ_{5g} modes and a coupling between the shear and the Γ_{5g} phason strains. This linear theory of elasticity, however, does not suggest any preference of the structural modulation with a particular symmetry. So we should go beyond the linear theory to discuss the energetic stability of the modulated phases.

Biham *et al.* [21] have derived some of the third order invariants to discuss the possible instabilities of the icosahedral structures. Here we make complete analyses in this line. We have 20 minimal couplings; 4 for the elastic strains, 5 for the phason strains and 11 for the phonon-phason coupling. The phonon-phason coupling terms contribute to inducing anisotropy in the elastic constants for the lower symmetric modulated phases, modifying the phason stiffness under an external hydrostatic pressure and so on. The third order invariants for the phason strains may cause instability of the icosahedral phase, which is assumed to be locally stable. Although a transition due to this instability is first order, it is still of interest to investigate possible lower symmetric phases as far as the transition is *weakly* first order.

The third order invariants for the phason strains are obtained with our definition of the normal modes in eqs.(4) and (5) as follows; one invariant from a triple product of the Γ_{4g} phason strains as

$$\mathcal{I}_4 = \frac{\sqrt{3}}{2}[f_1(f_1^2 - f_2^2 - f_3^2 - f_4^2) + 2\sqrt{5}f_2f_3f_4], \tag{12}$$

and two invariants from the Γ_{5g} ones as

$$\begin{aligned}\mathcal{I}_5 = \frac{1}{16}[\ &\sqrt{5}g_1(g_1^2 - 3g_2^2) + 3\sqrt{3}g_2(g_2^2 - 3g_1^2)\\ &- 12(\sqrt{5}g_1 + \sqrt{3}g_2)g_3^2 - 12(\frac{1}{\tau^2}g_1 - \sqrt{3}\tau g_2)g_4^2\\ &- 12(-\tau^2 g_1 + \sqrt{3}\frac{1}{\tau}g_2)g_5^2],\end{aligned} \tag{13a}$$

$$\begin{aligned}\mathcal{I}_5' = \frac{\sqrt{3}}{16}[\ &3\sqrt{3}g_1(g_1^2 - 3g_2^2) - \sqrt{5}g_2(g_2^2 - 3g_1^2) - 32g_3g_4g_5\\ &- 4(\sqrt{3}g_1 - \sqrt{5}g_2)g_3^2 - 4(-\sqrt{3}\tau g_1 - \frac{1}{\tau^2}g_2)g_4^2\\ &- 4(\sqrt{3}\frac{1}{\tau}g_1 + \tau^2 g_2)g_5^2].\end{aligned} \tag{13b}$$

Besides, we have two coupling terms between the Γ_{4g} and Γ_{5g} phason strains as

$$\begin{aligned}\mathcal{J} = &\frac{1}{2}g_1(f_2^2 - \tau f_3^2 + \frac{1}{\tau}f_4^2) + \frac{1}{2\sqrt{3}}g_2(-\sqrt{5}f_2^2 - \frac{1}{\tau^2}f_3^2 + \tau^2 f_4^2)\\ &+ \frac{1}{\sqrt{3}}g_3(\sqrt{5}f_1f_2 + f_3f_4) + \frac{1}{\sqrt{3}}g_4(\sqrt{5}f_1f_3 + f_2f_4)\\ &+ \frac{1}{\sqrt{3}}g_5(\sqrt{5}f_1f_4 + f_2f_3),\end{aligned} \tag{14a}$$

$$\begin{aligned}\mathcal{J}' = &\frac{1}{\sqrt{30}}f_1[3(g_1^2 + g_2^2) - 2(g_3^2 + g_4^2 + g_5^2)]\\ &+ \frac{1}{\sqrt{6}}f_2(\sqrt{3}g_1g_3 - \sqrt{5}g_2g_3 - 2g_4g_5) + \frac{1}{\sqrt{6}}f_3(-\sqrt{3}\tau g_1g_4 - \frac{1}{\tau^2}g_2g_4 - 2g_3g_5)\\ &+ \frac{1}{\sqrt{6}}f_4(\sqrt{3}\frac{1}{\tau}g_1g_5 + \tau^2 g_2g_5 - 2g_3g_4).\end{aligned} \tag{14b}$$

The phason free energy involving the nonlinear effect is given as

$$\begin{aligned}\mathcal{F} = &\frac{1}{2}A_4|\mathbf{f}|^2 + B_4\mathcal{I}_4 + \frac{1}{4}C_4|\mathbf{f}|^4\\ &+ \frac{1}{2}A_5|\mathbf{g}|^2 + B_5\mathcal{I}_5 + B_5'\mathcal{I}_5' + \frac{1}{4}C_5|\mathbf{g}|^4\\ &+ D_1\mathcal{J} + D_2\mathcal{J},\end{aligned} \tag{15}$$

where

$$|\mathbf{f}|^2 = \sum_{i=1}^{4} f_i^2, \qquad |\mathbf{g}|^2 = \sum_{i=1}^{5} g_i^2. \tag{16}$$

Here we include the trivial fourth order invariants with positive constants, C_4 and C_5, to ensure the global stability. We neglect 2 other nontrivial fourth order invariants and 8 fourth order couplings between the Γ_{4g} and Γ_{5g} phason strains. The local stability of the icosahedral phase requires positive stiffness, A_4 and A_5.

First let us neglect the third order couplings $\mathcal{J}$ and $\mathcal{J}'$ for simplicity. This corresponds to the case discussed by Biham *et al.*[21]. The third order invariants $\mathcal{I}_4$, $\mathcal{I}_5$

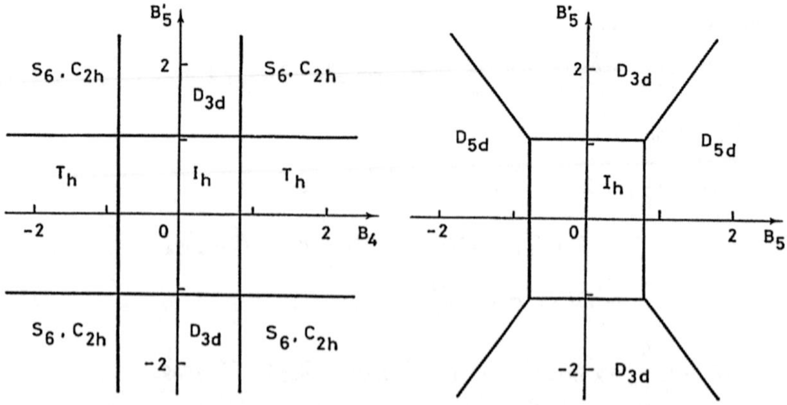

Fig.2: Phase diagrams for global minima of the free energy $\mathcal{F}$ with $D_1 = D_2 = 0$ in the $B_5 = 0$ (left) and $B_4 = 0$ (right) planes. A_4, C_4, A_5 and C_5 are taken as unity.

and $\mathcal{I}'_5$ are maximized for the phason strains in the T_h, D_{5d} and D_{3d} symmetric directions, respectively, with a fixed amplitude $|\mathbf{f}|$ and $|\mathbf{g}|$ as shown in Table 2, where $|\mathbf{f}|$ and $|\mathbf{g}|$ are taken as unity. The phason strains in the lower symmetric directions do not give any extrema of the third order invariants. Therefore the global minima of the free energy are found at the phason configurations giving the T_h symmetric modulation for $B_4^2 > 3/2A_4C_4$, the D_{5d} one for $B_5^2 > 5/8A_5C_5$ and $3|B_5| > \sqrt{5}|B'_5|$, and the D_{3d} one for $B_5'^2 > 9/8A_5C_5$ and $3|B_5| < \sqrt{5}|B'_5|$. If we have the Γ_{4g} phasons in the T_h symmetric direction and the Γ_{5g} phasons in the D_{3d} one, the symmetry of the global minimum state is given by the set of symmetry operations belonging to both of the T_h and D_{3d} point groups. Then we have eventually lower symmetric phases with the S_6 or C_{2h} symmetry. In Fig. 2, we show possible global minima of the free energy $\mathcal{F}$ in the parameter space spanned by three coupling constants B_4, B_5 and B'_5.

Within the linear theory of elasticity, we can treat the Γ_{4g} and Γ_{5g} phasons independently. Due to the third order couplings $\mathcal{J}$ and $\mathcal{J}'$, however, this is not the case within the nonlinear theory. Then we expect that the freezing of one phason mode induces the other phason. Suppose that the Γ_{5g} phasons are frozen in the D_{3d} direction, then we have

$$\mathcal{J} = \frac{g}{3}(2f_2^2 + \sqrt{5}f_1f_2 - f_3^2 - f_4^2 + f_3f_4),$$
$$\mathcal{J}' = \frac{4g^2}{3\sqrt{30}}(f_1 + \sqrt{5}f_2), \tag{17}$$

where $g = \pm|\mathbf{g}|$. Obviously $\mathcal{J}'$ causes the spontaneous deformation of the Γ_{4g} phason mode along the D_{3d} direction. On the other hand, since $\mathcal{J}$ is quadratic in f_i's, this coupling contributes to changing the stiffness of the Γ_{4g} phason mode. By diagonalizing the quadratic terms, we find that the effective phason stiffness constants are modified as $A_4 + 5/3D_1g$, $A_4 - 1/3D_1g$ (doubly degenerate) and $A_4 - D_1g$ with the principal axes along the D_{3d}, C_{2h} and S_6 symmetric directions, respectively. Depending on a sign of D_1g, the Γ_{4g} phason becomes soft in the D_{3d} or S_6 direction. This softening can make the spontaneous deformation much easier.

Fig. 3 shows the phase diagrams for the global minima of the free energy $\mathcal{F}$ with $D_1 = D_2 = 0.4$. By comparing with Fig. 2, we find that the higher symmetric T_h

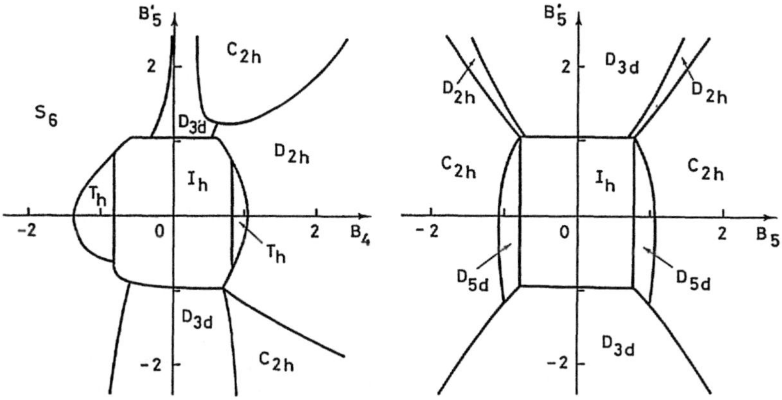

Fig.3: Similar figures to Fig.2 except for $D_1 = D_2 = 0.4$.

and D_{5d} phases are deformed into the lower symmetric ones; the T_h phase is deformed into the orthorhombic D_{2h} or the rhombohedral (trigonal) S_6 one while the pentagonal D_{5d} phase,into the monoclinic C_{2h} one. This is because of cooperative freezing of the phason strains caused by the third order couplings. The rhombohedral D_{3d} phase is, on the other hand, more stabilized by the cooperative freezing of the phason strains as discussed above.

Recently Widom [22] has conjectured the possibility of a structural phase transition caused by the linear phonon-phason coupling. The phason fields inducing such a transition should be the Γ_{5g} modes because the Γ_{4g} one is not coupled with the elastic (phonon) strains. In this sense this mechanism does not account for the symmetry change into the cubic phase, in which only the Γ_{4g} phason is frozen. However, Widom's elastic instability followed by the cooperative freezing of the Γ_{4g} phason strains will lead to a variety of modulated phases, especially the rhombohedral symmetry phases.

4. Concluding Remarks

The structural modulation induced by the phason degrees of freedom in icosahedral quasicrystals has been discussed. From the symmetry consideration, we have derived all the possible ways of structural modulation , which give the cubic, pentagonal, rhombohedral, orthorhombic and monoclinic symmetries. To discuss the energetic stability of these modulated structures, the elastic free energy involving the nonlinear terms has been derived.

There is a lot of experimental evidence of coexistence of the exotic crystalline and the icosahedral phases. In fact some of these crystalline phases are interpreted as the modulated quasicrystals discussed here. This implies that the instability against the structural modulation is inherent in some icosahedral quasicrystals. Then the third order terms in the phason strains determines which modulation is favorable. A recent conjecture of the phason softening inspired by discovery of a structural transition in AlFeCu quasicrystal [23,24] is of special interest in this context.

The stoichiometry is one of the most essential factors determining the stability of quasicrystals. In the 3d Penrose lattice, the stoichiometry is represented by the number ratio of the prolate and oblate rhombohedral units, which should be τ ideally. In the modulated phases, however, this ratio is different from the ideal value: in the orthorhombic phase in Fig.1, for example, it equals $84/52 = 1.615 \cdot \cdots$. Therefore the phenomenological constants in the phason free energy and so the frozen phason strains depend significantly on the sample composition. Systematic and well-controlled experiments in the vicinity of the ideal stoichiometry are surely desirable.

References

[1] P. J. Steinhardt and S. Ostlund, *The Physics of Quasicrystals*, (World Scientific, Singapore, 1987), Chap.1.

[2] P. Bak, Rept. Prog. Phys. **45**, 587 (1982).

[3] J. E. S. Socolar, T. C. Lubensky and P. J. Steinhardt, Phys. Rev. B**34**, 3345 (1986).

[4] D. M. Frenkel, C. L. Henley and E. D. Siggia, Phys. Rev. B**34**, 3649 (1986).

[5] D. Levine, T. C. Lubensky, S. Ostlund, S. Ramaswamy and P. J. Steinhardt, Phys. Rev. Lett. **54**, 1520 (1985).

[6] P. Bak, Phys. Rev. Lett. **54**, 1517 (1985); Phys. Rev. B**32**, 5764 (1985).

[7] P. A. Kalugin, A. Yu. Kitayev and L. S. Levitov, J. Phys. Lett. (Paris), **46**, L601 (1985).

[8] N. D. Mermin and S. Troian, Phys. Rev. Lett. **54**, 1524 (1985).

[9] C. L. Henley, J. Non-cryst. Solids, **75**, 91 (1985); this proceeding.

[10] K. H. Kuo, Mater. Sci. Forum, **22-24**, 131 (1987).

[11] V. Elser and C. L. Henley, Phys. Rev. Lett. **55**, 2883 (1985).

[12] Y. Ishii, J. Non-cryst. Solids, to be published.

[13] Y. Ishii , Phys. Rev. B**39**, 11862(1989).

[14] M. Hamermesh, *Group Theory* (Addison-Wesley, 1962), Chap.3.

[15] J. N. Pratt and G. V. Raynor, J. Inst. Met. **80**,449 (1952).

[16] J. E. S. Socolar and D. C. Wright, Phys. Rev. Lett. **59**, 221 (1987).

[17] R. N. Corby and P. J. Black, Acta. Cryst. B**33**, 3468 (1977).

[18] C. B. Shoemaker, D. A. Keszler and D. P. Shoemaker, Acta Cryst. B**45**, 13 (1989).

[19] L. Bendersky, Phys. Rev. Lett. **55**, 1461 (1985).

[20] M. Hamermesh, *Group Theory* (Addison-Wesley, 1962), Chap.5.

[21] O. Biham, D. Mukamel, and S. Strinkman, *Introduction to Quasicrystals.* edited by M. V. Jarić (Academic, Boston, 1988), Chap.5.

[22] M. Widom, *Proc. of the Adriatico Research Conference on Quasicrystals*, edited by M. V. Jarić and S. Lundqvist (World Scientific, Singapore,1990), pp. 337-355.

[23] A. I. Goldman, J. E. Shield, G. A. Guryan and P. W. Stephens, *Proc. of the Adriatico Research Conference on Quasicrystals*, edited by M. V. Jarić and S. Lundqvist (World Scientific, Singapore, 1990), pp. 67-73.

[24] P. A. Bancel, Phys. Rev. Lett. **63**, 2741 (1989); *ibid.* **64**, 496 (1990).

Part V

Electronic Structure

Electronic States in a Quasicrystal

T. Ninomiya

Department of Physics, University of Tokyo, Bunkyo-ku, Tokyo 113, Japan

Abstract. One-dimensional quasiperiodic lattice is regarded as a lattice obtained from a crystal by successive application of periodic modulation with the periods increasing as $n, n^2, n^3 \cdots$ (n is an integer). Formulation is given in the Wannier representation for the deformation of electronic states induced by the periodic modulations. In the inverse problem it is found that we can give the potential associated with the modulation so that the electronic states belonging to different bands have different characters. It is also shown that the characters of the wave functions can be different from those of the energy spectra.

1. Introduction

Fibonacci sequence of A and B is one of the typical quasiperiodic lattices in one dimension, and the electronic states on the lattice have been found to have various peculiar properties such as the self-similarity in the wave function of the edge state or the band center states and the singular continuous spectra of the energy eigenvalues. In the tight binding approach these critical properties have been found to be kept irrespective of the magnitude of the site or transfer energies assigned to A and B [1–6]. On the other hand, Harper equation

$$\psi_{n+1} + \psi_{n-1} + \lambda \cos(\sigma n + \theta)\psi_n = E\psi_n, \tag{1}$$

where $\sigma = \frac{(\sqrt{5}-1)}{2}$, gives the eigenstates which are dependent on the magnitude of λ : localized states for $\lambda > 2$, critical states for $\lambda = 2$, and extended states for $\lambda < 2$ [2]. Recently Hiramoto and Kohmoto showed that characters of eigenstates for other potentials are dependent on their energy eigenvalues [7].

Springer Series in Solid-State Sciences, Vol. 93 **Quasicrystals**
Editors: T. Fujiwara · T. Ogawa © Springer-Verlag Berlin, Heidelberg 1990

Thus, the electronic states on a quasiperiodic lattice show large varieties of their characters depending on details of Hamiltonian. The purpose of the present work is to discuss the electronic states on a quasiperiodic lattice (1D) in a way giving rather intuitive understanding of such peculiar properties and such large varieties. In Section 2 it is shown that a quasiperiodic lattice can be obtained from a crystal by successive application of periodic modulations, the periods increasing as $n, n^2, n^3 \cdots$, where n is an integer. Then, it will be reasonable to expect that we can have an intuitive picture of the electronic states, since the starting material is periodic and applied modulations are also periodic. Change of the wave function caused by each modulation will be formulated in Section 3 by Wannier representation. That is, we calculate the envelope function for each modulation and expect that a manner of change of the envelope function with successive modulation will determine the character of the wave function. In Section 4 we consider the inverse problem : that is, we will give a recipe for obtaining the periodic potential, which is non-local in general, responsible for given characters of wave functions and energy spectra. It is found that different bands can have different characters, and, furthermore, the character of wave function can be given independently from that of the energy spectra of its band.

2. Structural Features of Quasicrystals

Fibonacci sequence of A and B is normally generated by repeating the transformation

$$A \rightarrow AB \quad \text{and} \quad B \rightarrow A. \tag{2}$$

If we denote the sequence obtained after l-th generation by S_l, we have

$$S_l = S_{l-1}S_{l-2} = S_{l-2}S_{l-3}S_{l-2}. \tag{3}$$

Let us define X_m and Y_m by

$$X_m = X_{m-1}Y_{m-1}X_{m-1}, \tag{4}$$

and

$$Y_m = X_{m-1}Y_{m-1}, \tag{5}$$

with $X_0 = ABA$ and $Y_0 = AB$; then we have

$$X_n = S_{2m+2}, \tag{6}$$

and

$$Y_m = S_{2m+1}. \tag{7}$$

We now consider a method of obtaining X_s from a crystal by giving periodic modulations successively [4]. We start with the crystal

$$X_0X_0X_0X_0X_0X_0 \cdots\cdots X_0X_0X_0, \tag{8}$$

which contains $3^s X_0$. If we give the first modulation by replacing X_0 with Y_0 for every three X_0, we obtain $X_1X_1X_1 \cdots\cdots X_1$ which consists of $3^{s-1}X$:

$$\begin{aligned} &X_0X_0X_0X_0X_0X_0 \cdots\cdots X_0X_0X_0 \\ &\longrightarrow X_0Y_0X_0X_0Y_0X_0 \cdots\cdots X_0Y_0X_0 = X_1X_1 \cdots\cdots X_1. \end{aligned} \tag{9}$$

Second modulation is to replace X_1 with Y_1 for every three X_1 :

$$\begin{aligned} &X_1X_1X_1X_1X_1X_1 \cdots\cdots X_1X_1X_1 \\ &\longrightarrow X_1Y_1X_1X_1Y_1X_1 \cdots\cdots X_1Y_1X_1 = X_2X_2 \cdots\cdots X_2(3^{s-2}X_2). \end{aligned} \tag{10}$$

By repeating similar modulations we have

$$X_mX_m \cdots\cdots X_m(3^{s-m}X_m), \tag{11}$$

after m-th modulation, and we finally have X_s as

$$X_{s-1}X_{s-1}X_{s-1} \longrightarrow X_{s-1}Y_{s-1}X_{s-1} = X_s. \tag{12}$$

In this method the periods increase by a factor of three at every step of the modulation and the modulations are self-similar. Self-similarity is not always requested for quasiperiodicity, and by keeping only the rule of the increase of periodicity we can generalize the quasiperiodic lattice :

A one-dimensional quasiperiodic lattice is obtained from a crystal by applying periodic modulations successively, where the periods increase as $n, n^2, n^3, \cdots$ (n =integer). Here, the non-periodicity of an infinitely long quasiperiodic lattice is a result of infinite times of modulations, and systematic dependence of the modulation on m will determine the character of the eigenstates.

3 Electronic States on a Quasiperiodic Lattice : Wannier Representation

In this section we derive the change of wave functions induced by the periodic modulations. Since the lattice after the $(m-1)$-th modulation has the period n^{m-1}, the range of the wave number k_{m-1} is given by

$$-\frac{\pi}{n^{m-1}} \leq k_{m-1} < \frac{\pi}{n^{m-1}}. \tag{13}$$

When the m-th modulation is applied, the k-space is folded and the range is given by

$$-\frac{\pi}{n^{m}} \leq k_{m} < \frac{\pi}{n^{m}}, \tag{14}$$

and each band is split into n bands. Thus, the total number of the bands is n^m. Then, the assignment of the eigenstates is done by α_m for the band and k_m for the wave number. After the final modulation (the s-th modulation) with the periodic boundary condition, the only possible value of k_s is zero and each band contains only one eigenstate.

Let us consider the Schroedinger equation after the m-th modulation,

$$\begin{aligned} H_m\psi_m(\alpha_m,k_m:r) &= H_{m-1}\psi_m(\alpha_m,k_m:r)+\sum_{r'}V_m(r,r')\psi_m(\alpha_m k_m:r') \\ &= E_m(\alpha_m,k_m)\psi_m(\alpha_m,k_m:r) \end{aligned} \tag{15}$$

where $V_m(r,r')$ is the non-local potential associated with the m-th modulation. Wannier function $A_m(\alpha_m, l_m : r)$ corresponding to the wave function $\psi_m(\alpha_m,k_m:r)$ is defined by

$$\psi_m(\alpha_m,k_m:r) = \frac{1}{\sqrt{N_m}}\sum_{l_m}\exp(ik_m l_m n^m)A_m(\alpha_m,l_m:r), \tag{16}$$

where l_m denotes the lattice site corresponding to the lattice expressed as $X_mX_m\cdots X_m$, which consists of $n^{s-m}X_m$ and $N_m = n^{s-m}$. Since the set of Wannier functions makes a complete set, ψ_m can be expressed by a linear combination of $A_{m-1}(\alpha_{m-1},l_{m-1}:r)$, which are defined in a way similar to eq(16), as

$$\psi_m(\alpha_m,k_m:r) = \sum_{\alpha_{m-1}}\sum_{l_{m-1}} f_m(\alpha_m,k_m:\alpha_{m-1},l_{m-1})A_{m-1}(\alpha_{m-1},l_{m-1}:r). \tag{17}$$

The envelope function f_m gives a deformation of the wave function due to the m-th modulation. From a textbook on solid state physics we find the equation for $f_m(\alpha_m, k_m : \alpha_{m-1}, l_{m-1})$ as

$$\sum_{l'_{m-1}} \varepsilon_{m-1}(\alpha_{m-1}, l_{m-1} - l'_{m-1}) f_m(\alpha_m, k_m : \alpha_{m-1}, l'_{m-1})$$
$$+ \sum_{\alpha'_{m-1}} \sum_{l'_{m-1}} \widetilde{V}_m(\alpha_{m-1}, l_{m-1} : \alpha'_{m-1}, l'_{m-1}) f_m(\alpha_m, k_m : \alpha'_{m-1}, l'_{m-1})$$
$$= E_m(\alpha_m, k_m) f_m(\alpha_m, k_m : \alpha_{m-1}, l_{m-1}), \tag{18}$$

where $\widetilde{V}_m$ is defined by

$$\widetilde{V}_m(\alpha_{m-1}, l_{m-1} : \alpha'_{m-1}, l'_{m-1})$$
$$= \sum_r \sum_{r'} A^*_{m-1}(\alpha_{m-1}, l_{m-1} : r) V_m(r, r') A_{m-1}(\alpha'_{m-1}, l'_{m-1} : r'), \tag{19}$$

and satisfies

$$\widetilde{V}_m(\alpha_{m-1}, l_{m-1} + n : \alpha'_{m-1}, l'_{n-1} + n) = \widetilde{V}_m(\alpha_{m-1}, l_{m-1} : \alpha'_{m-1}, l'_{m-1}) \; .$$

The inverse transformation of eq(19) is

$$V_m(r, r') = \sum_{\substack{\alpha_{m-1} \\ l_{m-1}}} \sum_{\substack{\alpha'_{m-1} \\ l'_{m-1}}} A_{m-1}(\alpha_{m-1}, l_{m-1} : r)$$
$$\times \widetilde{V}_m(\alpha_{m-1}, l_{m-1} : \alpha'_{m-1}, l'_{m-1}) A^*_{m-1}(\alpha'_{m-1}, l_{m-1} : r'). \tag{20}$$

As seen in eq(18), $\widetilde{V}_m(\alpha_{m-1}, l_{m-1} : \alpha'_{m-1}, l'_{m-1})$ gives mixing of the bands α_{m-1} and α'_{m-1} induced by the application of m-th modulation $V_m(r, r')$. If we rewrite eq(20) as

$$V_m(r, r') = \sum_{\alpha_{m-1}} \sum_{\alpha'_{m-1}} V_m(\alpha_{m-1}, \alpha'_{m-1} : r, r'), \tag{21}$$

$V_m(\alpha_{m-1}, \alpha'_{m-1} : r, r')$ gives the component of $V_m(r, r')$ which contributes to the band mixing. ε_{m-1} is defined by

$$\varepsilon_{m-1}(\alpha_{m-1}, l_{m-1} - l'_{m-1})$$
$$= \frac{1}{\sqrt{N_{m-1}}} \sum_{k_{m-1}} E_{m-1}(\alpha_{m-1}, k_{m-1}) \exp i k_{m-1}(l_{m-1} - l'_{m-1}). \tag{22}$$

If we define a second kind of Wannier function, $a_m(\alpha_m, l_m : \alpha_{m-1}, l_{m-1})$, by

$$f_m(\alpha_m, k_m : \alpha_{m-1}, l_{m-1}) = \frac{1}{\sqrt{N_m}} \sum_{l_m} \exp(ik_m l_m n^m) a_m(\alpha_m, l_m : \alpha_{m-1}, l_{m-1}), \qquad (23)$$

we obtain the following equation :

$$\begin{aligned} A_m(\alpha_m, l_m : r) &= \sum_{\substack{\alpha_{m-1} \\ l_{m-1}}} \sum_{\substack{\alpha_{m-2} \\ l_{m-2}}} \cdots\cdots \sum_{\substack{\alpha_0 \\ l_0}} a_m(\alpha_m, l_m : \alpha_{m-1}, l_{m-1}) \\ &\times a_{m-1}(\alpha_{m-1}, l_{m-1} : \alpha_{m-2}, l_{m-2}) \cdots a_1(\alpha_1, l_1 : \alpha_0, l_0) \\ &\times A_0(\alpha_0, l_0 : r). \end{aligned} \qquad (24)$$

Let us discuss the case of no band mixing, that is, the case of $\tilde{V}_m(\alpha_{m-1}, l_{m-1} : \alpha'_{m-1}, l'_{m-1}) = 0$ for $\alpha_{m-1} \neq \alpha'_{m-1}$ in more detail for discussions of the inverse problem given in the next section.

a) Before application of the m-th modulation :

Before the m-th modulation the wave function is specified with α_{m-1} and k_{m-1}, where the region of k_{m-1} is given by

$$-\frac{\pi}{n^{m-1}} \leq k_{m-1} < \frac{\pi}{n^{m-1}} \ . \qquad (13')$$

We here write the wave function before the m-th modulation by the envelope function in the reduced zone scheme corresponding to the m-th modulation : that is, we use $k_m(-\pi/n^m \leq k_m < \pi/n^m)$ instead of k_{m-1} and consider that the band α_{m-1} consists of n bands $(\alpha^j_{m-1} : j = 0, 1, 2, \cdots n-1)$. The envelope function f^0_m is

$$f^0_m(\alpha^j_{m-1}, k_m : \alpha_{m-1}, l_{m-1}) = \frac{1}{\sqrt{N_{m-1}}} \exp(ik^j_m l_{m-1} n^{m-1}), \qquad (25)$$

where k^j_m is defined by

$$k_m > 0$$

$$j = \text{even} \ : k^j_m = k_m + \frac{2\pi}{n^m} \times \frac{j}{2}$$

$$= \text{odd} \quad : k_m^j = k_m - \frac{2\pi}{n^m} \times \frac{j+1}{2}$$

$$k_m < 0 \tag{26}$$

$$j = \text{even} \quad : k_m^j = k_m - \frac{2\pi}{n^m} \times \frac{j}{2}$$

$$= \text{odd} \quad : k_m^j = k_m + \frac{2\pi}{n^m} \times \frac{j+1}{2}$$

and

$$\sum_{l'_{m-1}} \varepsilon_{m-1}(\alpha_{m-1}, l_{m-1} - l'_{m-1}) \exp(ik_m^j l'_{m-1} n^{m-1}) = E_{m-1}(\alpha_{m-1}, k_m^j) \exp(ik_m^j l_{m-1} n^{m-1}). \tag{27}$$

b) After application of the m-th modulation :

The m-th modulation $V_m(\alpha_{m-1}, l_{m-1} : \alpha_{m-1}, l_{m'-1})$ produces n eigenstates $(s = 0, 1, \cdots n-1)$ for each k_m by mixing of the plane waves (α_{m-1}^j, k_m) $(j = 0, 1, 2, \cdots n-1)$ given by eq(25) :

$$f_m(\alpha_m = \alpha_{m-1}^s, k_m : \alpha_{m-1}, l_{m-1}) = \sum_j P_j(s, k_m) \exp(ik_m^j l_{m-1} n^{m-1}). \tag{28}$$

The equations for $P_j(s, k_m)$ are given by

$$\sum_{j'} \overset{\approx}{V}_m(\alpha_{m-1}, j; \alpha_{m-1}, j') P_{j'}(s, k_m) = [E_m(\alpha_{m-1}^s, k_m) - E_{m-1}(\alpha_{m-1}, k_m^j)] P_j(s, k_m) \tag{29}$$

where

$$\overset{\approx}{V}_m(\alpha_{m-1}, j; \alpha_{m-1}, j') = \frac{1}{N_m} \sum_{l_{m-1}} \sum_{l'_{m-1}} \exp(-ik_m^j l_{m-1} n^{m-1}) \times \tilde{V}_m(\alpha_{m-1}, l_{m-1} : \alpha_{m-1}, l'_{m-1}) \exp(ik_m^{j'} l'_{m-1} n^{m-1}) \ . \tag{30}$$

Equations (29) are the secular equations to be solved, but we now change our stand-point from solving the secular equations to the inverse problem.

4. Inverse Problem

In the inverse problem we look for the potential which is responsible for given characters of wavefunctions and energy spectra. To be more explicit, we give prescribed characters to $P_j(s,k_m), E_m(\alpha^s_{m-1},k_m)$ and $E_{m-1}(\alpha_{m-1}k^j_m)$ in eq(29). Then, we have n equations ($s = 0,1,2,\cdots n-1$) to be solved simultaneously for obtaining $\overset{\approx}{V}_m(\alpha_{m-1},j;\alpha_{m-1},j')$ where $j' = 0,1,2,\cdots n-1$ and j is fixed. We have n sets of ($j = 0,1,2,\cdots n-1$) of simultaneous equations to give all the components of the potential ($j,j' = 0,1,2,\cdots,n-1$) .

In order that eq.(29) gives the unique solution, the determinant of the coefficients $P_j(s,k_m)$ should be non-zero :

$$\begin{vmatrix} P_0(0,k_m) & P_1(0,k_m) & \cdots & P_{n-1}(0,k_m) \\ P_0(1,k_m) & P_1(1,k_m) & \cdots & P_{n-1}(1,k_m) \\ & & \cdots & \\ P_0(n-1,k_m) & P_1(n-1,k_m) & \cdots & P_{n-1}(n-1,k_m) \end{vmatrix} \neq 0 \,. \qquad (31)$$

This is satisfied, since the wave functions are given to be orthogonal to each other.

By assuming an appropriate dependence on the generation m for the wave function and the energy spectra, we can give arbitrary characters to them. For example, if the envelope functions associated with each modulation are not dependent on m and the energy spectra are self-similar, we shall obtain critical states. Another interesting case may be following : let us consider the case that a certain band obtained by m_1-th modulation remains unchanged for further modulation. That is, for this band $V_m(\alpha_{m-1},\alpha_{m-1}:r,r') = 0$ for $m > m_1$. Then, electronic states belonging to this band have the wave functions corresponding to the periodicity n^{m_1} , although the total potential $V_m(r,r')$ has the larger periodicity n^m, and can be regarded as extended states. If the electronic states in the adjacent band are localized or critical, we have a "mobility edge" at the boundary of these two bands. Strictly speaking the boundary is an energy gap, but if $m_1 >> 1$, the gap width may be negligible compared with the original band width and actually we have the mobility edge. We can also see that in eq.(29) we can give a character to the energy spectra in a manner uncorrelated with that of the wave function.

These considerations lead to the conclusions :

(1) On the one-dimensional quasiperiodic lattice different bands can have different characters,

(2) The characters of the wave functions can be different from those of the energy spectra.

The same way as that developed for the one-dimensional case will be possible for two-or three-dimensional cases, because the grids dual to 2-D (or 3-D) Penrose lattice are obtained by applying periodic modulations successively in the x- and y- directions (x-, y- and z-directions) [8].

References

[1] M. Kohmoto, L.P. Kadanoff and C. Tang : Phys. Rev. Lett. **50,** 1870 (1983)

[2] S. Ostlund, R. Pandit, D. Rand, H.J. Schellnhuber and E.D. Siggia, Phys. Rev. Lett. **50,** 1873 (1983)

[3] M. Fujita and K. Machida, Solid State Commun. **59,** 61 (1986)

[4] T. Ninomiya, J. Phys. Soc. Jpn. **55,** 3709 (1986)

[5] K. Machida and M. Fujita, Phys. Rev. **B34,** 7367 (1986)

[6] M. Kohmoto, B. Sutherland and C. Tang, Phys. Rev. **B35,** 1020 (1987)

[7] H. Hiramoto and M. Kohmoto, Phys. Rev. **B40,** 8225 (1989)

[8] T. Ninomiya : J. Noncrystalline Solids, Proceedings of the 7-th Conf. on Liquid and Amorphous Metals (Kyoto, 1989)

Scaling of Structures, Spectra and Wave Functions

T. Janssen

Institute for Theoretical Physics, University of Nijmegen,
6525 ED Nijmegen, The Netherlands

Scaling invariance of quasiperiodic tilings is discussed and spectra of phonons and electrons in well-ordered but aperiodic structures are analyzed. The latter may show self-similarity even if the underlying structures are not scale invariant.

1. Introduction

When quasicrystals were discovered it was soon realized that there was possibly a connection between their structure and quasiperiodic tilings. As for the Penrose tiling in two dimensions these tilings can sometimes be constructed using deflation and inflation rules. These reflect a special property of such tilings: scale invariance. On the other hand one has found that spectra for phonons and electrons in quasiperiodic systems sometimes show self-similarity. Such self-similarity was found by Hofstadter for Harper's equation[1], by de Lange and Janssen for the modulated spring model[2] and the modulated Kronig-Penney model[3], and later on by several people for the Fibonacci chain. It is clear that not all of these structures have scaling properties themselves. For example, the sinusoidally modulated chain does not have this property. Nevertheless, we want to address here the question whether there is a relation between scale invariance of a structure and the character of its phonon or electron spectrum.

First we shall discuss the possible scale transformations for one-dimensional systems, and a method to find them for higher-dimensional systems. Then a number of models for phonons and electrons in one-dimensional quasi-periodic systems will be discussed, and an analysis of the multifractal properties of their spectra will be given. These are compared with the properties for systems that have also long-range order, but are not quasi-periodic. As examples the Cantor and Thue-Morse chains will be discussed.

2. Scaling Structures in One- and Higher- dimensional Spaces

Consider a quasiperiodic tiling of rank n in a d-dimensional space V_E for which the vertices are the intersections of V_E with atomic surfaces in the lattice points of an n-dimensional lattice. Then each vertex is a linear combination with integral coefficients of n basis vectors. A scale transformation transforms

Springer Series in Solid-State Sciences, Vol. 93 Quasicrystals
Editors: T. Fujiwara · T. Ogawa © Springer-Verlag Berlin, Heidelberg 1990

each such basis vector into a linear integral combination of the basis vectors, i.e. one has

$$S\vec{a}_i = \sum_{j=1}^{n} \Gamma(S)_{ji}\vec{a}_j. \qquad (1)$$

In this case a basis transformation maps a vertex of the n-dimensional lattice that has a certain distance to the physical space V_E to a vertex that has generally another distance to V_E. If this distance is smaller, one has the situation that a lattice point in n dimensions that gives rise to a vertex of the tiling is mapped to another lattice point also giving a vertex of the tiling.

Consider the case of a rank two tiling in one-dimensional space. A basis transformation is an element of GL(2,Z). It has integer trace p and determinant +1 or -1.

If the determinant is +1 the eigenvalues satisfy $\lambda + \lambda^{-1} = p$. Therefore, λ is one out of an infinite series:

$$\lambda = (p - \sqrt{(p^2 - 4)})/2, \quad \text{with } \mid p \mid > 2. \qquad (2)$$

For a basis (a, b), $(\lambda a, \lambda^{-1} b)$ of the lattice the physical space is an eigenspace belonging to the eigenvalue λ. When the atomic surface is a line of length $b(1+\lambda^{-1})$ each vertex of the tiling goes under multiplication with a factor λ to another vertex.

If the determinant is -1 the corresponding eigenvalues belong to an infinite series given by

$$\lambda = (\sqrt{(p^2 + 4)} - p)/2, \quad \text{with } p \neq 0. \qquad (3)$$

In both cases one may choose the value of b in such a way that the two-dimensional lattice is square. The irrationality of the structure, corresponding to the orientation of the lattice with respect to V_E, is given by $\alpha = \sqrt{\lambda}$ for the Det=+1 case and $\alpha = \lambda$ for the Det=-1 case.

For higher-dimensional tilings one may determine the existence of a scale invariance in the following way[4]. The embedding of the quasiperiodic structure into the n-dimensional space is (partially) fixed by the action of the point group K on the Fourier module. If R is an element of K its action on basis vectors is given by

$$R\vec{a_i}^* = \Gamma(R)^*_{ji}\vec{a_j}^*. \qquad (4)$$

The matrices $\Gamma(R)^*$ form a representation of K, the contragredient of which is reducible into at least two components: D_E and D_I. For convenience we assume here that these two representations do not have irreducible components in common: $(D_E, D_I) = 0$.

If Γ(S) is an element of the *centralizer* of $\Gamma(K)$ one has $\Gamma(S)\Gamma(R)\Gamma(S)^{-1} = \Gamma(R)$. If in addition both D_E and D_I are irreducible, it follows from Schur's lemma that $\Gamma(S)$ is equivalent with the direct sum of a multiple of the identity in V_E and a multiple of the identity in V_I. It follows that this determines a scaling transformation, because $\mid Det(\Gamma(S)) \mid = 1$. As an example consider an octagonal two-dimensional tiling of rank 4. The generators of the point group

K are

$$\Gamma(\alpha) = \begin{pmatrix} 0 & 0 & 0 & -1 \\ 1 & 0 & 0 & 0 \\ 0 & 1 & 0 & 0 \\ 0 & 0 & 1 & 0 \end{pmatrix} , \quad \Gamma(\beta) = \begin{pmatrix} 0 & 0 & 0 & 1 \\ 0 & 0 & 1 & 0 \\ 0 & 1 & 0 & 0 \\ 1 & 0 & 0 & 0 \end{pmatrix} . \tag{5}$$

An element of the centralizer is

$$\Gamma(S) = \begin{pmatrix} 1 & 1 & 0 & -1 \\ 1 & 1 & 1 & 0 \\ 0 & 1 & 1 & 1 \\ -1 & 0 & 1 & 1 \end{pmatrix} , \tag{6}$$

which has 2 eigenvalues $(1+\sqrt{2})$ and 2 eigenvalues $(1\text{-}\sqrt{2})$. The eigenspace for the first two is V_E, from which follows the scaling property.

A slightly more complicated example is the scaling for a sevenfold rank six tiling in two dimensions (cf.[5]). An element of the centralizer of the point group generated by the matrices

$$\Gamma(\alpha) = \begin{pmatrix} 0 & 0 & 0 & 0 & 0 & -1 \\ 1 & 0 & 0 & 0 & 0 & -1 \\ 0 & 1 & 0 & 0 & 0 & -1 \\ 0 & 0 & 1 & 0 & 0 & -1 \\ 0 & 0 & 0 & 1 & 0 & -1 \\ 0 & 0 & 0 & 0 & 1 & -1 \end{pmatrix} \quad \text{and} \quad \Gamma(\beta) = \begin{pmatrix} 0 & 0 & 0 & 0 & 0 & 1 \\ 0 & 0 & 0 & 0 & 1 & 0 \\ 0 & 0 & 0 & 1 & 0 & 0 \\ 0 & 0 & 1 & 0 & 0 & 0 \\ 0 & 1 & 0 & 0 & 0 & 0 \\ 1 & 0 & 0 & 0 & 0 & 0 \end{pmatrix} \tag{7}$$

is the element

$$\Gamma(S) = \begin{pmatrix} 0 & -1 & 0 & 0 & 0 & 1 \\ 0 & -1 & -1 & 0 & 0 & 1 \\ 1 & -1 & -1 & -1 & 0 & 1 \\ 1 & 0 & -1 & -1 & -1 & 1 \\ 1 & 0 & 0 & -1 & -1 & 0 \\ 1 & 0 & 0 & 0 & -1 & 0 \end{pmatrix} , \tag{8}$$

which has double eigenvalues -2(cos($\pi/7$)+cos($3\pi/7$)), 2(cos($2\pi/7$)-cos($\pi/7$)) and 2(cos($2\pi/7$)-cos($3\pi/7$)). Therefore, this 7-fold tiling has a scaling factor of -2.24698.

A second type of scaling transformation is obtained from the *normalizer*. If S belongs to the normalizer one has $S\Gamma(R)S^{-1} = \Gamma(\pi R)$, where π is an automorphism of K. Again it follows from Schur's lemma that V_E is invariant under S, if $(D_E, D_I) = 0$ and $D_E \circ \pi$ is equivalent to the irreducible representation D_E. The requirement that the automorphism does not change the equivalence class of the representation is essential here, and is related to the requirement one has to demand for the equivalence of superspace groups. Consider, as an example, a 2-dimensional rank four dodecagonal tiling with point group K generated by

$$\Gamma(\alpha) = \begin{pmatrix} 0 & 0 & 0 & -1 \\ 1 & 0 & 0 & 0 \\ 0 & 1 & 0 & 1 \\ 0 & 0 & 1 & 0 \end{pmatrix}, \quad \Gamma(\beta) = \begin{pmatrix} 0 & 0 & 0 & 1 \\ 0 & 0 & 1 & 0 \\ 0 & 1 & 0 & 0 \\ 1 & 0 & 0 & 0 \end{pmatrix}. \tag{9}$$

The two matrices

$$T_1 = \begin{pmatrix} 1 & 0 & 0 & -1 \\ 1 & 1 & 0 & 0 \\ 0 & 1 & 1 & 1 \\ 0 & 0 & 1 & 1 \end{pmatrix} \quad \text{and} \quad T_2 = \begin{pmatrix} 1 & -1 & 1 & 0 \\ 1 & -1 & 0 & 0 \\ 0 & 1 & -1 & 1 \\ -1 & 1 & -1 & 1 \end{pmatrix} \tag{10}$$

both belong to the normalizer. The first maps α to α and β to $\alpha^{-1}\beta$ and the corresponding π leaves the equivalence class of D_E invariant, because it leaves the conjugacy classes invariant. It has eigenvalues $\lambda = \sqrt{2+\sqrt{3}}\exp(\pm\pi/12)$ and acts in V_E as a combination of a scale factor and a rotation over $\pi/12$. The second matrix maps α to α^5 and β to $\alpha^{-1}\beta$, and interchanges V_E and V_I. Therefore, the latter does not lead to a scale transformation.

A third type of scale transformation is obtained from an *intertwining operator* that does not belong to the normalizer: a matrix S not in GL(n,Z) such that for every $R \in K$ one has $S\Gamma(R) = \Gamma(\pi R)S$, again with the condition that π leaves the equivalence class of D_E invariant. An example is the matrix

$$S = \begin{pmatrix} 1 & 0 & 0 & 0 & 1 \\ 1 & 1 & 0 & 0 & 0 \\ 0 & 1 & 1 & 0 & 0 \\ 0 & 0 & 1 & 1 & 0 \\ 0 & 0 & 0 & 1 & 1 \end{pmatrix}, \tag{11}$$

which is an intertwining operator for the point group of the Penrose tiling generated by

$$\Gamma(\alpha) = \begin{pmatrix} 0 & 0 & 0 & 0 & 1 \\ 1 & 0 & 0 & 0 & 0 \\ 0 & 1 & 0 & 0 & 0 \\ 0 & 0 & 1 & 0 & 0 \\ 0 & 0 & 0 & 1 & 0 \end{pmatrix} \quad \text{and} \quad \Gamma(\beta) = \begin{pmatrix} 0 & 0 & 0 & 0 & 1 \\ 0 & 0 & 0 & 1 & 0 \\ 0 & 0 & 1 & 0 & 0 \\ 0 & 1 & 0 & 0 & 0 \\ 1 & 0 & 0 & 0 & 0 \end{pmatrix}. \tag{12}$$

In this case the automorphism π is the identity. The matrix S does not belong to GL(5,Z) because its determinant is 2. However, it gives a scaling transformation in V_E with factor $-\tau$ (τ is the golden number) and corresponds to the inflation of the Penrose tiling.

The existence of scaling invariance immediately leads to a corresponding transformation of the tiling. If a tiling consists of p types of tiles, the scaling transformation with factor λ gives a tiling with larger tiles, which can be subdivided into tiles from the original set. An example is given in Figure 1 for a 2-dimensional rank four octagonal tiling, composed of squares and rhombuses

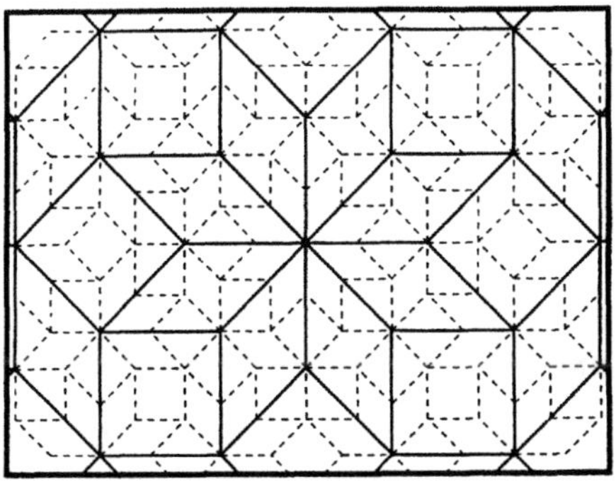

Figure 1: Scaling of the octagonal tiling

with an angle of 45°. Scaling with a factor $1+\sqrt{2}$ gives another tiling (thick lines), the elements of which can be subdivided: a big square consists of six half squares and four rhombuses, a big rhombus consists of four half squares and three rhombuses. Therefore, if N_s indicates the number of squares, and N_r the number of rhombuses, the scale transformation gives a linear transformation

$$\begin{pmatrix} N_r \\ N_s \end{pmatrix} \rightarrow \begin{pmatrix} 3 & 2 \\ 4 & 3 \end{pmatrix} \begin{pmatrix} N_r \\ N_s \end{pmatrix}, \tag{13}$$

on the set of tiles. The eigenvalues of the transformation matrix are $(1 \pm \sqrt{2})^2$, which are the squares of the eigenvalues of the scale transformation matrix.

3. Scaling Properties of Spectra

In a number of quasiperiodic model systems one has found self-similarity in the spectra of phonons and electrons. Hofstadter calculated the spectra for a series of commensurate values of the magnetic field in the case of Harper's equation for a Bloch electron in a magnetic field[1]. Similar results were reported for the sinusoidally modulated spring model, a linear chain with modulated spring constants[2], and for the modulated Kronig-Penney model[3], where δ functions are located on the sites of a displacively modulated chain. In this case self-similarity occurs in the energy-wave vector plane. A further analysis has been given for spectra of a tight-binding model for the case of a Fibonacci chain[6], which has wave vector τ-1.

One way to analyze scaling properties of a spectrum uses a thermodynamic approach[7]. A cruder idea of scaling properties is obtained when looking at the total spectral measure as a function of the step in the commensurate approximation, or at the width of one special band.

In the thermodynamic formalism one approximates the irrational wave vector γ by a series of rational approximants $\gamma_p = L_p/N_p$. One calculates for each step p the partition function, free energy and entropy, respectively, as

$$Z_p(\beta) = \sum_{i=1}^{N_p} \Delta_i^{\beta} \; ; \; F_p(\beta) = log(Z_p(\beta))/p \; ; \; S_p(\epsilon) = F_p(\beta) - \beta \frac{\mathrm{dF}}{\mathrm{d}\beta} \, . \tag{14}$$

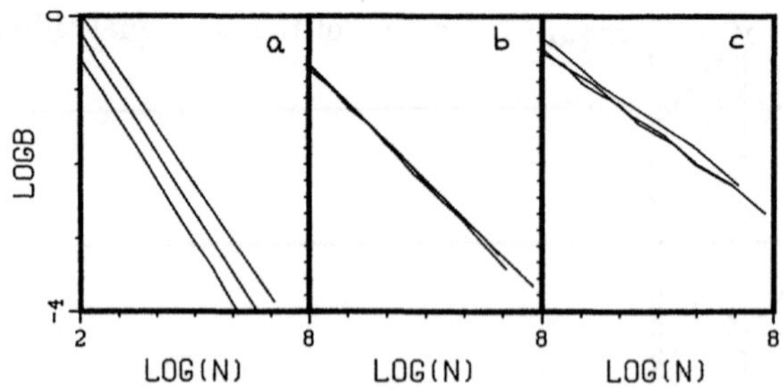

Figure 2: Total spectrum (a) and top (b) and middle (c) band widths for approximants with period N=N_p to quasiperiodic systems. Quasiperiodicities as in Eq.(3) with p=1,2,3

If the limit for p tending to infinity of $S_p(\epsilon)$ exists it gives the distribution of scaling exponents:the number of bands with scaling exponents between ϵ and $\epsilon + \delta\epsilon$ is proportional to $\exp(nS(\epsilon))$.

The modulated spring model is a harmonic chain with spring constant α_n between the n-th and (n-1)-th atom. The quasiperiodic function α_n can be chosen as $\alpha(1\text{-}\Delta\ \cos(2\pi\gamma \mathrm{n}+\phi))$. For Δ=1 the logarithm of the total spectral measure is a linear function of the step p, where N_p is the p-th Fibonacci number[8]. The functions $S_p(\epsilon)$ converge to a smooth entropy function, which is an indication for a singular continuous spectrum.

For comparing the situations for various quasiperiodic systems the total spectral measure and the widths of top and middle bands are given in Figure 2 for a tight-binding model with discrete Schrödinger equation

$$t_{n+1}c_{n+1} \ + \ t_n c_{n-1} \ = \ E \ c_n, \tag{15}$$

where the hopping frequencies t_n form a quasiperiodic rank two sequence. For sequences with irrationality as in eqs. 2 and 3 there is a non-trivial scaling (p=1,2 and 3). For other non-scaling structures with a quadratic irrational wave vector the continued fraction expansion is periodic and if one takes these periods as steps, there is again scaling: the logarithm of the band width tends to a linear function of the logarithm of the unit cell length. For a number of non-quadratic irrational numbers the scaling is not perfect. Even for very large unit cells there are some zigzags in the functional dependence. However, approximately, there is again scaling and the scale exponents are comparable with those of the scaling structures. Generally, the (average) scale exponent of a quasiperiodic structure seems to be a smooth function of the irrationality γ. The maximum of this function lies at γ=1/2.

To look for a deviation of this behaviour we look at systems that are also well ordered, but not quasiperiodic. The first example is the *Cantor chain.* Consider an alphabet with two letters, A and B. A substitution π is defined as π(A)=ABA and π(B)=BBB. Then, starting from A, one gets after the p-th iteration a chain of length 3^p. The infinite sequence is not quasiperiodic.

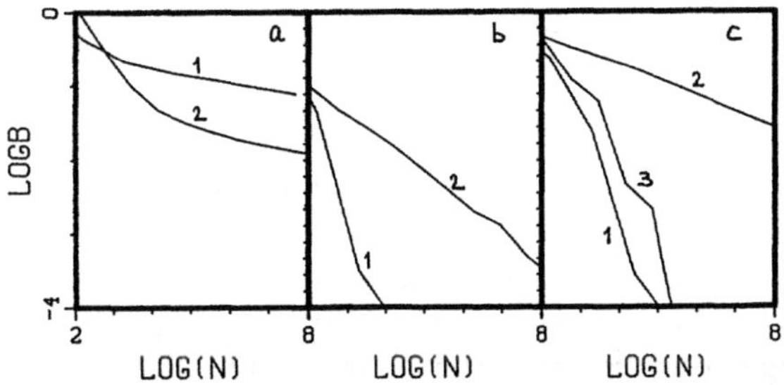

Figure 3: Total spectral measure (a), and widths of top (b) and middle (c) bands for approximants of the Cantor (1) and Thue-Morse (2) chains. With the middle bands is given a gap band of the Thue-Morse chain (3).

The corresponding tight-binding Schrödinger equation then can be solved by a transfer matrix technique. The transfer matrix is determined by the expression

$$\begin{pmatrix} c_n \\ c_{n-1} \end{pmatrix} \Longrightarrow \begin{pmatrix} c_{n+1} \\ c_n \end{pmatrix} = T_n \begin{pmatrix} c_n \\ c_{n-1} \end{pmatrix}. \tag{16}$$

The matrices T_n depend on the energy E. Then one can say that E belongs to the spectrum if $|Tr(\prod_n T_n(E)| \leq 2$. The trace of the product of 3^p matrices can be found from those of $A = T_0$, $B = T_1$ and T_0T_1, which are a,b and c, respectively. Then $\mathrm{Tr}(\pi A) = ac - b$, $\mathrm{Tr}(\pi B) = b^3 - 3b$, and $\mathrm{Tr}(\pi AB) = -abc + b^2c^2 - b^2 - c^2 + 2$. Hence the trace can be found from the nonlinear mapping of three-dimensional space

$$(a, b, c) \rightarrow (ac - b, b^3 - 3b, b^2c^2 - b^2 - c^2 - abc + 2). \tag{17}$$

Using this nonlinear mapping one finds the spectrum. The dependence of spectral measure and widths of top and middle bands on the approximation is given in Figure 3. The behaviour is very different from that of the quasiperiodic systems. For increasing length the spectral measure goes down. However, the quantity

$$\frac{B_p - B_{p+1}}{B_{p+1} - B_{p+2}} \tag{18}$$

goes to unity, which is an indication that the spectrum does not go exponentially to a constant. The width of the top band goes faster to zero than exponentially, which is an indication of localized states, and the same holds for the band in the middle of the spectrum.

Another system that is well ordered, but not quasiperiodic is the *Thue-Morse chain*[9]. It is obtained in a similar way via the substitutions π(A)=AB and π(B)=BA. Therefore one has $N_p = 2^p$. In this case the nonlinear map is

$$(a, b, c) \rightarrow (c, c, 2 - a^2 - b^2 + abc), \tag{19}$$

This is essentially a two-dimensional mapping, because after one iteration a point is mapped into the plane $x_1 = x_2$. There it corresponds, after change of variables, to the mapping

$$(x, y) \quad \rightarrow \quad (4(x-1)y^2 \; + \; 1, \; x) \tag{20}$$

studied by Axel et al.[9]. This 2-dimensional map has a fixed point (x,y) = (1,1) corresponding to a trace equal to two. The whole line $x = 1$ is mapped by (20) into the fixed point. The pre-image of this line is another line. Repeating this procedure one finds a fractal sheaf of lines which are finally mapped to the fixed point. The corresponding states are characterized by a wave vector in the centre of the Brillouin zone. However, not all states are attracted to the fixed point.

If one calculates the energy spectrum by means of this nonlinear map one finds a behaviour that can be compared to that of the Cantor chain[10, 11, 12]. Again the spectral measure does not go exponentially to a limit. The spectral measure is decreasing for increasing lengths of the unit cell, but numerically it is very hard to decide whether the spectral measure in the limit vanishes or remains finite. The top and middle bands both scale, but in a trivial way with exponents 1/2 and 1, respectively. The former value corresponds to edge states. The spectrum in the p-th approximation, with unit cell equal to $N_p = 2^p$, consists of N_p bands which pairwise touch. The states in these points are doubly degenerate. There are $N_p/2 - 1$ open gaps. If one follows a band just above such a gap, its width goes faster than exponentially to zero. This is an indication for localized states. On the other hand the degenerate states, which correspond to the points in the mapping that are finally mapped to the fixed point, are extended.

This behaviour can also be seen from a calculation of the entropy function $S(\epsilon)$. For the Cantor as well as the Thue-Morse case there is a limit for this function. In both cases there is a point at infinity, which points to the existence of localized states. The points for finite values of ϵ correspond to extended, and possibly critical, states. The entropy functions are given in Figure 4. The Hausdorff dimension of the spectrum is the slope of the tangent from the origin to the $S(\epsilon)$ curve. For both the Cantor and the Thue-Morse chain, the slope is smaller than unity. Its precise value depends on the ratio of the hopping frequencies over intervals A and B. This is in agreement with [12].

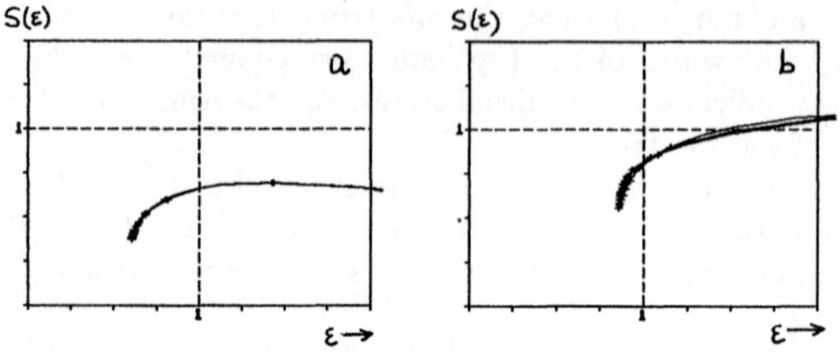

Figure 4: Entropy functions for the spectra of the Cantor (a) and Thue-Morse (b) chain

The spectral properties of higher-dimensional quasiperiodic structures are less well known. Preliminary calculations in 2 and 3 dimensions point to the existence of localized states as well as extended states. The density of states shows many (probably an infinity of) peaks. The analysis of the spectral properties still has to be given.

Acknowledgement

The author has benefitted much from stimulating discussions with Aloysio Janner, André Katz and Mahito Kohmoto. Thanks are due to the Dutch Organization for Fundamental Research of Matter (FOM) for partial support.

References

[1] D.Hofstadter, Phys.Rev. **B14**, 2239 (1976)

[2] C.de Lange and T.Janssen, J.Phys. C**14**,5269 (1981)

[3] C.de Lange and T.Janssen, Phys.Rev. B**28**,195 (1983)

[4] A.Janner and T.Janssen, Proceedings 3rd International Conference on Quasicrystals, Mexico 1989.

[5] K.Niizeki, J.Phys. C**22**,193 (1989)

[6] M.Kohmoto, L.P.Kadanoff and Chao Tang, Phys.Rev.Lett. **50**, 1870 (1983)

[7] M.Kohmoto, Phys.Rev.A **37**, 1345 (1988)

[8] T.Janssen and M.Kohmoto, Phys.Rev.B **38**,5811 (1988)

[9] F.Axel, J.P.Allouche, M.Kléman and M.Mendès-France, J.de Physique, Colloques **7**, C3-181 (1986)

[10] Z.Cheng, R.Savit and R.Merlin, Phys.Rev.B **37**, 4375 (1988)

[11] G.C.La Rocca, Sol.St.Comm. **70**, 115 (1989)

[12] F.Axel and J.Peyriere, preprint (1989)

Multifractal Method for Spectra and Wave Functions of Quasiperiodic Systems

M. Kohmoto

Institute for Solid State Physics, University of Tokyo,
7-22-1 Roppongi, Minato-ku, Tokyo 106, Japan

Abstract. A quasiperiodic Schrödinger equation in one dimension can have an energy spectrum which is absolutely continuous, singular continuous, or a point spectrum. The corresponding wave functions are extended, critical, or localized, respectively. The multifractal method of analyzing the energy spectrum and the wave functions is presented. This is shown to be effective in distinguishing the three types of wave functions. Some examples are briefly discussed.

1. Introduction

A one-dimensional quasiperiodic tight-binding model (or discrete Schrodinger equation) is written as

$$-\psi_{n+1} - \psi_n + \lambda V(n\omega)\psi_n = E\psi_n\,, \tag{1.1}$$

where ω is an irrational number and $V(x)$ is a periodic function, i.e., $V(x+1) = V(x)$. When the potential part is random, it represents the Anderson model of localization. The random electronic (or phonon) system is known to be always localized in one dimension (rigorous) and two dimensions. Therefore one has to go to three dimensions to study the transition between the localized and the extended states[1].

It is known that the quasiperiodic systems can have localized states[2]. In fact, they can have both extended and localized states even in one dimension, where one can apply many analytical or numerical techniques.

The quasiperiodic models studied intensively are: (A) $V(x) = \cos(2\pi x)$[3], (B) $V(n\omega)$ having two values 1 and -1 which are arranged in a Fibonacci sequence[4,5]. In the model (A), all the states are extended for $\lambda < 2$ (purely absolutely continuous spectrum) and all the states are localized for $\lambda > 2$ (purely dense point spectrum). At the critical coupling $\lambda_c = 2$, all the states are critical (purely singular continuous spectrum). All the states are critical for model (B) irrespective of λ. This was recently proven rigorously by Suto[6] following the work of Kotani[7].

The model (A) is sometimes called the Harper equation. Two-dimensional electrons in a uniform magnetic field with a periodic potential can be described by this equation. The magnetic flux per unit cell is ω and λ represents the anisotropy (critical coupling $\lambda_c = 2$ corresponds to the isotropic case)[8]. Spectra of the Harper equation are shown in Fig. 1.

For a rational value of $\omega = p/q$ (where p and q are integers which are mutually prime), the spectrum consists of q bands. On the other hand, we have a Cantor set for an irrational ω, i.e. gaps are dense in the spectrum. Although the spectra in Fig. 1 are all Cantor sets for an irrational ω, they are seemingly different. For $\lambda = 2$, the Lebesgue measure of the spectrum is zero and the fractal dimension is less than one,

Springer Series in Solid-State Sciences, Vol. 93 Quasicrystals
Editors: T. Fujiwara · T. Ogawa

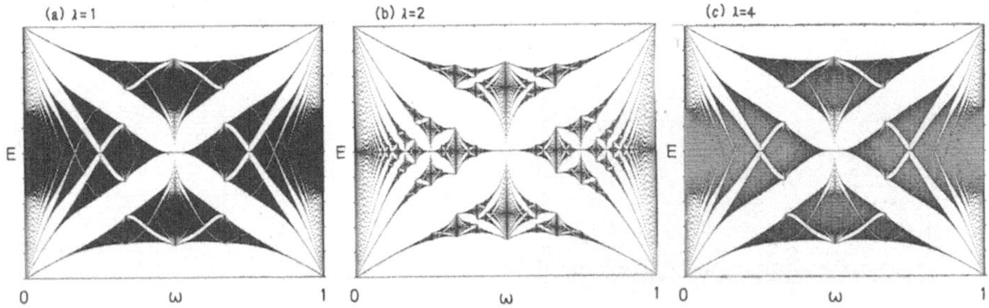

Fig. 1 Spectra of the Harper equation. The energy is normalized appropriately.

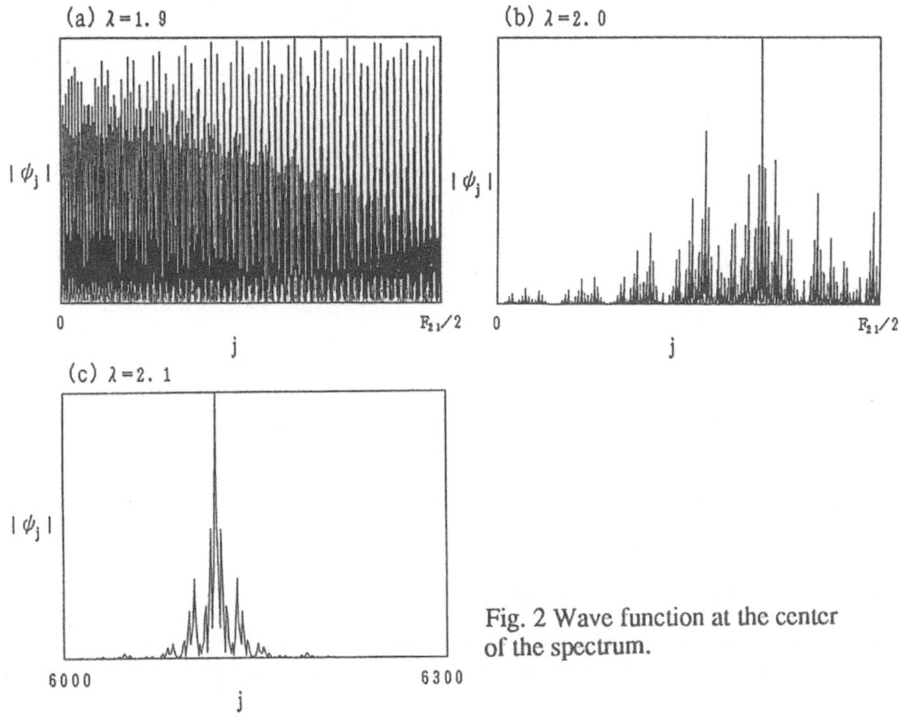

Fig. 2 Wave function at the center of the spectrum.

while the Lebesgue measure is nonzero and the fractal dimension is one for $\lambda < 2$. The spectrum for $\lambda > 2$ is dense-point and its closure is the same as that for $\lambda' = \lambda/4 < 2$ due to the duality[3].

The wave functions at the center and the edge of the spectrum are shown in Figs. 2 and 3.

The three types of wave functions are distinct in the figures. It is, however, important to characterize the wave functions qualitatively.

An analysis of a fractal with a probability measure (multifractal) through a certain partition function was put forward by Halsey *et al.* [9]. We adopt the multifractal method to analyze the spectra and wave functions of the quasiperiodic systems[10].

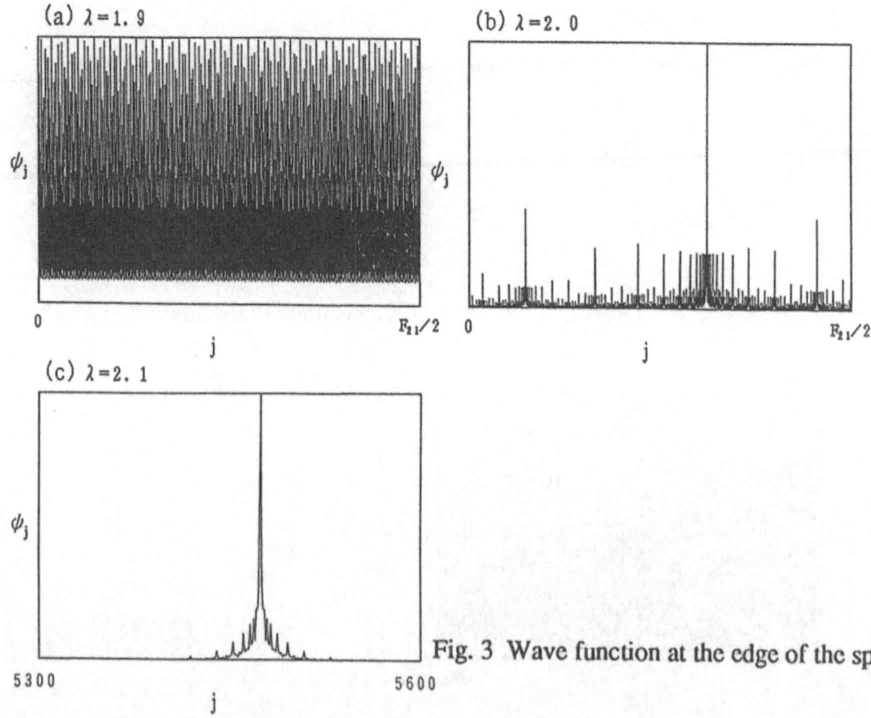

Fig. 3 Wave function at the edge of the spectrum.

Although it seems that similar ideas have been developed by other groups, I review here mostly the work of my collaborators and myself without putting it in perspective with other work. This is due to my inability to follow the literature closely and I apologize for this.

2. Scaling Analysis of the Spectra and Wave Functions

In a scaling analysis of a fractal set with measure (multifractal), we consider a systematic partition of the set. The nth level of the partition consists of a number of bars with length l_i. A scaling index for l_i is given by $l_i \sim e^{-n\varepsilon_i}$. We consider a situation where a probability measure p_i is associated with each bar. The scaling index of singularity is given by $p_i \sim l_i^{\alpha_i}$. Now distributions of ε and α specify the scaling properties of the multifractal. Here one can use a formalism equivalent to statistical mechanics to obtain the distributions. The analysis of the spectrum is a special case where we have a distribution of l_i, but where p_i is constant. On the other hand, the wave function has a distribution of p_i, but l_i is constant. For a more detailed discussion on the statistical-mechanics of multifractals, see Ref. 11.

2.1 Spectrum

In order to understand the scaling of the spectrum, we need to define appropriate scaling indices and the entropy function for them. It is convenient then to consider systematic approximations or finite partitions of the Cantor-set spectrum. This can be done by replacing the irrational number ω by a series of rational numbers which are obtained by truncating the continued-fraction expansion of ω. For example, the inverse of the golden mean $[\tau^{-1} = (\sqrt{5}-1)/2]$ is approximated by a series of rational numbers $F_{n-1}/F_n = \{\frac{1}{2}, \frac{2}{3}, \frac{3}{5}, \frac{5}{8}, \frac{8}{13}, ...\}$ where F_n is a Fibonacci number defined recursively as $F_{-1} = F_0 = 1$ and $F_{n+1} = F_n + F_{n-1}$. At the nth level of approximation of the Cantor-set spectrum, we have N bands whose widths are denoted by $\Delta_i (i = 1, ... ,N)$. The number of bands grows exponentially with respect to n as $N \sim a^n$ (in the example above $N = F_n$ and $a = \tau = (\sqrt{5}+1)/2$).

Let us define a scaling index for Δ_i by

$$\varepsilon_i = -\frac{1}{n} \ln\Delta_i . \tag{2.1}$$

We also define an entropy function $S(\varepsilon)$ by

$$S(\varepsilon) = \frac{1}{n} \ln \Omega(\varepsilon) , \tag{2.2}$$

where $\Omega(\varepsilon)d\varepsilon$ is the number of bands whose scaling index lies between ε and $\varepsilon + d\varepsilon$. Here it is important to notice that Δ_i and $\Omega(\varepsilon)$ depend exponentially on n. A band at the nth level splits into many bands at a higher level and may thus yield a number of different values of the scaling indices ε. However, we expect that the entropy function which represents the distribution of ε will converge to a smooth limiting form as n tends to infinity, and give the complete information about the scaling behavior.

As in the formalism of statistical mechanics, it is convenient to introduce a "partition function" and a "free energy," which are defined by

$$Z_n(\beta) = \sum_{i=1}^{N} \Delta_i^{\beta} , \tag{2.3}$$

and

$$F(\beta) = \lim_{n\to\infty} \frac{1}{n} \ln Z_n(\beta) . \tag{2.4}$$

Once the free energy is calculated, the entropy function is obtained by a Legendre transformation,

$$S(\varepsilon) = F(\beta) + \beta\varepsilon , \tag{2.5}$$

$$\varepsilon = -\frac{dF(\beta)}{d\beta} . \tag{2.6}$$

Thus by changing "temperature" β one can pick a value of ε and then the corresponding $S(\varepsilon)$ is calculated. On the other hand, β can be written in terms of ε as

$$\beta = \frac{dS(\varepsilon)}{d\varepsilon} . \tag{2.7}$$

Usually $S(\varepsilon)$ is defined on an interval $[\varepsilon_{min}, \varepsilon_{max}]$ and there is no scaling behavior corresponding to ε which is outside the interval and $S(\varepsilon) = 0$. However, $F(\beta)$ is still defined there and from (2.5) it is given by $F(\beta) = -\varepsilon_{max}\beta$ for $\beta > \beta_{min}$ and $F(\beta) = -\varepsilon_{min}\beta$ for $\beta < \beta_{max}$. Thus useful information is only contained in $F(\beta)$ for the region between β_{min} and β_{max} where it is not linear.

The Hausdorff dimension D_H is the zero β_c of the free energy [see (2.3) and (2.4)],i.e.,

$$F(D_H) = 0 . \tag{2.8}$$

From (2.5), (2.7), and (2.8) we have

$$\beta_c = \frac{S(\varepsilon)}{\varepsilon} = \frac{dS(\varepsilon)}{d\varepsilon} . \tag{2.9}$$

Thus the Hausdorff dimension $D_H = \beta_c$ has a geometrical interpretation in ε - $S(\varepsilon)$ plot: it is a slope of the line through the origin tangent to $S(\varepsilon)$.

The index ε which represents scaling of the Lebesgue measure of the energy spectrum can actually be related to the singular behavior of the density of states. At the nth level of approximation, each band carries the same number of states: $p_i = 1/N = a^{-n}$ (the total number of states is normalized to unity, i.e., $\sum p_i = 1$). An index α_i which represents the singular behavior of the density of states is defined as

$$p_i \sim \Delta_i^{\alpha_i} . \tag{2.10}$$

Since $p_i = 1/N = a^{-n}$ and $\Delta_i \sim e^{-n\varepsilon_i}$ [see (2.1)], it is related to ε by

$$\alpha\varepsilon = \ln a. \tag{2.11}$$

The spectrum of singularity introduced by Halsey *et al.* [9] is given by

$$\Omega'(\alpha) \sim \langle\Delta\rangle^{-f(\alpha)} , \tag{2.12}$$

where $\Omega'(\alpha)d\alpha$ is a number of bands whose scaling index α lies between α and $\alpha+d\alpha$, namely $\Omega'(\alpha) = \Omega(\varepsilon)\,|d\varepsilon/d\alpha|$, and $\langle\Delta\rangle$ is a representative value of Δ which was not specified clearly in Ref. 9. If one identifies $\langle\Delta\rangle = \exp(-n\varepsilon)$[see (2.1)], $f(\alpha)$ can be related to the entropy function by

$$f(\alpha) = \frac{S(\varepsilon)}{\varepsilon} . \tag{2.13}$$

This simple relation between $f(\alpha)$ and $S(\varepsilon)$ holds since the measure p_i is constant in this case. In general the measure has its own scaling behavior and we do not have a simple relation like (2.13).

An absolute continuous spectrum, for which the states are extended, has a nonsingular density of states (apart from possible van Hove singularities) and α is given by 1 [see (2.10)]. On the other hand, a point spectrum, which corresponds to localized states, would give $\alpha = 0$. If α is different from 0 or 1, the spectrum has a non-trivial scaling and probably one can expect a singular continuous spectrum. The corresponding wave functions are neither localized nor extended in the standard way and are called "critical." Thus the entropy functions $S(\varepsilon)$ and $f(\alpha)$ give the essential information on the spectral type and the nature of the wave function.

2.2. Wave Function

A wave function is defined on lattice sites which are regular. Therefore the wave function is not singular nor fractal as it is. However, for the incommensurate problem there is a consistent way to take a scaling limit of the lattice which is a continuous interval [0,1]. Then the wave function defined on the interval can have singularities and scaling which we shall analyze.

In the nth approximation the system is periodic with period N. Consider now the square of the wave function at site i as a probability measure, namely,

$$p_i = |\psi_i|^2 . \tag{2.14}$$

We normalize p_i by

$$\sum_{i=1}^{N} p_i = \sum_{i=1}^{N} |\psi_i|^2 = 1 , \tag{2.15}$$

for a finite system with N sites. Assign a uniform Lebesgue measure $l = N^{-1}$ to all the sites. Then the size of the system is normalized to unity. In the limit of n tending to infinity, the probability measure is defined on the interval [0,1] and one can discuss singularities and scaling. Here the support of the probability measure is non-fractal, but the distribution of the measure can have scaling. Note the difference from the previous case of the spectrum in which the probability measure is uniformly assigned but the support may be a Cantor set or a fractal.

The scaling index for the Lebesgue measure is given by $l = \exp(-n\varepsilon)$ and is constant:

$$\varepsilon = \ln a . \tag{2.16}$$

The scaling index for the probability measure is defined as

$$p_i = l^{\alpha_i} , \tag{2.17}$$

and the entropy for α is defined as

$$S'(\alpha) = \frac{1}{n} \ln\Omega(\alpha) , \tag{2.18}$$

where $\Omega(\alpha)d\alpha$ is the number of sites which have index between α and $\alpha + d\alpha$. As n is increased a single site becomes many sites and it is not possible in general to follow a single scaling index. However we expect that the entropy function $S'(\alpha)$ which represents the distribution of α converges to a smooth limiting form as n tends to infinity. Thus the scaling behavior of the wave function is well represented by $S'(\alpha)$.

As in the previous case of the spectrum, it is convenient to introduce a partition function

$$Z'(q) = \sum_i p_i^q , \tag{2.19}$$

and free energy

$$G(q) = \frac{1}{n} \ln Z'(q) . \tag{2.20}$$

The entropy function is given by the Legendre transformation

$$S'(\alpha) = G(q) + q\alpha\varepsilon , \tag{2.21}$$

and

$$\alpha\varepsilon = -\frac{dG(q)}{dq} . \tag{2.22}$$

The function $f(\alpha)$ in this case is defined as[9]

$$\Omega(\alpha) \sim 1^{-f(\alpha)} , \tag{2.23}$$

and is related to the entropy function by [see (2.18)]

$$S'(\alpha) = \varepsilon f(\alpha). \tag{2.24}$$

For an extended wave function we should have $\alpha = 1$, and a localized wave function has $f(\alpha)$ consisting of two points: $f = 0$ at $\alpha = 0$ and $f = 1$ at α going to infinity. The point $\alpha = 0$ corresponds to the sites with nonzero Ψ_i and the point α at infinity all the other sites. For a critical wave function with a distribution of scaling indices we expect to have a smooth $f(\alpha)$ defined in a finite region $[\alpha_{min}, \alpha_{max}]$.

3. Examples of Multifractal Analysis

3.1. Fibonacci Model

The off-diagonal version of the Fibonacci model is written as

$$t_{j+1}\psi_{j+1} + t_j\psi_{j-1} = E\psi_j , \tag{3.1}$$

where t_j's (j=1,2,....) take two values t_A and t_B arranged in a Fibonacci sequence T_∞ which is defined as the limit of a recursion $T_{n+1} = T_nT_{n-1}$ with $T_1 = \{A\}$ and $T_2 = \{AB\}$. So one has $T_3 = \{ABA\}$, $T_4 = \{ABAAB\}$, and so on. In order to investigate the infinite system, we take a series of finite systems T_n whose number of sites is a Fibonacci number F_n defined by $F_{n+1}=F_{n-1} + F_n$ with $F_0 = F_1 = 1$.

A multifractal analysis of the spectrum of this model was given by Kohmoto, Sutherland, and Tang[12]. Here we report an exact $f(\alpha)$ for the wave function at the center of the spectrum which was obtained by Fujiwara, Kohmoto, and Tokihiro[13]

In order to investigate the wave functions on the Fibonacci lattice, (3.1) is written as $\Psi_j = M(t_{j+1},t_j)\Psi_{j-1}$ with $\Psi_j = \begin{bmatrix}\psi_{j+1}\\ \psi_j\end{bmatrix}$, $\Psi_{j-1} = \begin{bmatrix}\psi_j\\ \psi_{j-1}\end{bmatrix}$ and a transfer matrix $M(t_{j+1},t_j) = \begin{bmatrix}E/t_{j+1} & -t_j/t_{j+1}\\ 1 & 0\end{bmatrix}$. A wave function is obtained by multiplying Ψ_0 by transfer matrices successively once the energy is taken in the spectrum. There are three types of transfer matrices $M(t_A,t_A)$, $M(t_A,t_B)$, and $M(t_B,t_A)$ (note that there are no successive B's in the Fibonacci sequence). Denote the product of the first F_n transfer matrices by $M^{(n)}$, then we have a renormalization-group (RG) map

$$M^{(n+1)} = M^{(n-1)}M^{(n)}, \tag{3.2}$$

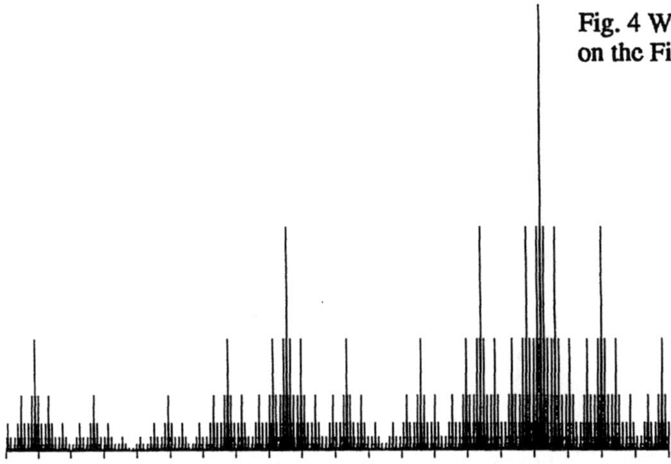

Fig. 4 Wave function at E = 0 on the Fibonacci lattice.

with $M^{(1)} = B = M(t_A,t_A)$ and $M^{(2)} = A = M(t_A,t_B)M(t_B,t_A)$[4,5,13]. Let us define $x_n = \mathrm{Trace}\, M^{(n)}/2$, then the energy spectrum of a periodic system consisting of T_n's is determined by a condition $|x_n| < 1$. It can be shown[4] that x_n obeys a trace map

$$x_{n+1} = 2x_n x_{n-1} - x_{n-2}. \tag{3.3}$$

The energy E enters in the initial condition and those which give bounded orbits determine the spectrum. In particular, the center of the spectrum has a six cycle[14] and the outer most edge of the spectrum has a two cycle[11]. Most of the bounded orbits , however, are chaotic.

When E = 0, not only the trace map (3.3) but also the full RG map (3.2) has a six cycle starting from $B=M(t_A,t_A)=\begin{bmatrix} 0 & 1 \\ -1 & 0 \end{bmatrix}$, and $A=M(t_A,t_B)M(t_B,t_A)=\begin{bmatrix} -R & 0 \\ 0 & 1/R \end{bmatrix}$ with $R = t_B/t_A$[13]. The wave function is shown in Fig. 4.

This remarkable property allows us to determine $f(\alpha)$ exactly as[13]

$$f(\alpha) = \frac{1}{3\ln\tau}\,[\ln y(R^{2q}) - q\frac{d}{dq}\ln y(R^{2q})] \tag{3.4}$$

and

$$\alpha = \frac{1}{3\ln\tau}\,[\ln y(R^2) - \frac{d}{dq}\ln y(R^{2q})], \tag{3.5}$$

with

$$y(x) = (1/2x)\,\{(x+1)^2 + \sqrt{(x+1)^4 + 4x^2}\,\}, \tag{3.6}$$

where τ is the golden mean $(\sqrt{5}+1)/2$. It is plotted in Fig. 5. The range of α where $f(\alpha)$ is defined is given by $[\alpha_{min}, \alpha_{max}] = [1/(3\ln\tau)\{\ln y(R^2) - \ln R^2\}, 1/(3\ln\tau)\{\ln y(R^2) + \ln R^2\}]$, and $f(\alpha) = 0$ for both α_{min} and α_{max}. The minimum value α_{min} corresponds to the largest square amplitude of the wave function and α_{max} corresponds to the smallest one. The maximum value of $f(\alpha)$ gives the Hausdorff

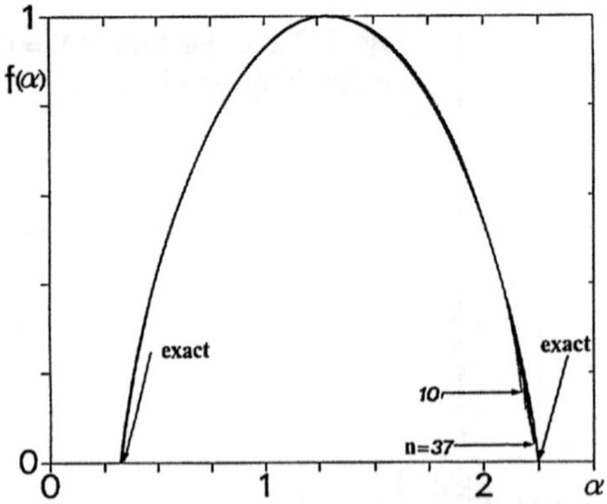

Fig. 5 The exact f(α) together with those calculated for finite systems.

dimension of the support of the wave function[9]. This is one as it should be since the support is an interval [0,1] in our formulation. The maximum occurs at $\alpha = \ln y(R^2)/(3\ln\tau)$. It is not difficult to prove that f(α) is symmetric about the maximum as is seen in Fig.5.

3.2. Harper Equation

The spectrum was analyzed by Tang and Kohmoto[16]. At the critical point $\lambda = 2$, the spectrum is singular continuous and it has a smooth f(α). In the subcritical region $\lambda < 2$, f(α) consists of two points: f =1 for α = 1 and f = 0 for α = 1/2. The point f = 1 for α = 1 represents that the spectrum is absolutely continuous with a non-zero Lebesgue measure (i.e. fractal dimension is one). The point f = 0 for α = 1/2 represents the remnants of the Van Hove singularities of the periodic systems. In the supercritical region $\lambda > 2$, f(α) does not converge well as the system size is increased. This is a sign of a point spectrum which has localized wave functions.

The analysis of the wave functions is given in Hiramoto and Kohmoto[17]. Since a finite system always gives a smooth f(α) irrespective of the nature of the wave functions, it is very important to do an extrapolation from finite-size data. It turns out that the calculation of f(α) at α_{min} which represents the largest $|\psi_j|^2$'s is most effective in distinguishing the localized, critical, and extended wave functions. The wave function at the outermost edge of the spectrum which is shown in Fig.2 has been analyzed and the result is shown in Fig. 6
The following behaviors are clearly seen:

$$f(\alpha_{min}) = \begin{cases} 1, & \alpha_{min} = 1 & \text{for the extended wave function ,} \\ 0, & \alpha_{min} \neq 0,\ 1 & \text{for the critical wave function,} \\ 0, & \alpha_{min} = 0 & \text{for the localized wave function.} \end{cases}$$

The leading finite-size correction is O(1/n). This is consistent with the formal analogy between the present method and statistical mechanics in canonical ensemble formalism (see eqs. (2.19) and (2.20); n corresponds to the system size). At the critical point $\lambda = 2$, f(α) is determined to be a smooth curve and shown in Fig.7.

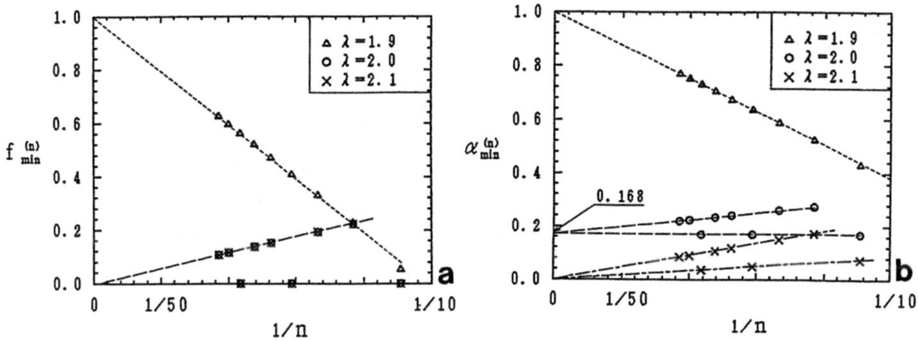

Fig. 6 Determination of (a) f_{min} and (b) α_{min}.

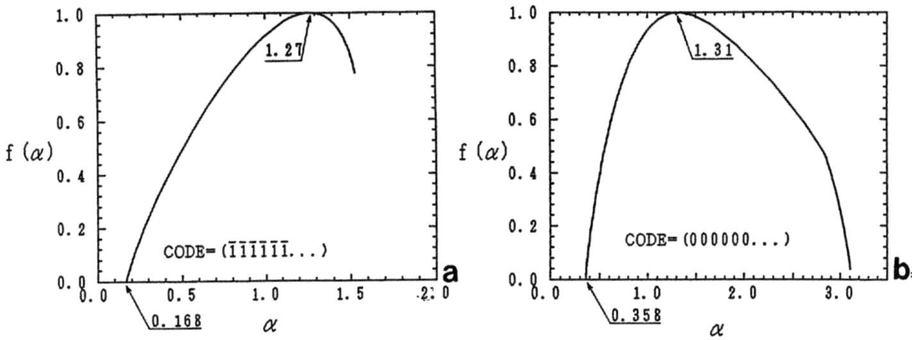

Fig. 7 f(α) for the wave functions for the Harper equation with λ = 2 at (a) the edge, and (b) the center of the spectrum.

References

1) See e.g. Lee,P., and Ramakrishnan,T.V., Rev. Mod. Phys. **57**, 287 (1985).

2) See, for example, Simon,B., Adv. Appl. Math. **3**, 463 (1982); Sokoloff,J.B., Phys. Rep. **126**, 189 (1985).

3) Aubry,S. and Andre,G., Ann. Isr. Phys. Soc. **3**, 133 (1980).

4) M. Kohmoto, L.P. Kadanoff, and C.Tang, Phys. Rev. Lett. **50**, 1870 (1983).

5) S. Ostlund, R. Pandit, D. Rand, H.J. Schellnhuber, and E.D. Siggia, Phys. Rev. Lett. **50**, 1873 (1983).

6) A. Suto, "Singular Continuous Spectrum on a Cantor Set of Zero Lebesgue Measure for the Fibonacci Hamiltonian", to appear in J. Stat. Phys.

7) S. Kotani, "Jacobi Matrices with Random Potentials Taking Finitely Many Values" Preprint, University of Tokyo (1988).

8) P.G. Harper, Proc. Phys. Soc. Lond. A**68**, 874 (1955); D.R. Hofstadter, Phys. Rev. B**14**, 2239 (1976); M. Kohmoto, Phys.Rev.B**39**, 11943 (1989).

9) T.C. Halsey, M.H. Jensen, L.P. Kadanoff, I. Procaccia, and B.I. Shraiman, Phys. Rev. A**33**,1141 (1986).

10) T. Janssen and M. Kohmoto, Phys. Rev. B38, 5811 (1988).
11) M. Kohmoto, Phys. Rev. A37, 1345 (1988).
12) M. Kohmoto, B. Sutherland, and C. Tang, Phys. Rev. B35, 1024 (1987).
13) T. Fujiwara, M. Kohmoto, and T. Tokihiro, Phys. Rev. B40, 7413 (1989).
14) M. Kohmoto and J.R. Banavar, Phys. Rev. B34, 563 (1986).
15) M. Kohmoto, and Y. Oono, Phys Lett. 102A, 745 (1985).
16) C. Tang and M. Kohmoto, Phys. Rev. B34, 2041 (1986).
17) H. Hiramoto and M. Kohmoto, Phys. Rev. B40, 8225 (1989).

Localization in One-Dimensional Quasiperiodic Systems

H. Hiramoto

Institute of Materials Science, University of Tsukuba, Tsukuba, Ibaraki 305, Japan

Abstract. Electronic properties in one-dimensional quasiperiodic (QP) systems are studied by using a scaling analysis. First, the one-electron problem in a QP potential is studied, and it is shown that the mixed spectrum with mobility edges can appear even in one-dimension. Next, the effect of an electron-electron interaction (U) in the Harper model and the Fibonacci model is studied within the Hartree-Fock (HF) approximation. In the Fibonacci model, the HF one-body spectrum remains singular continuous for small U, while in the Harper model with the critical case, the singular continuous spectrum disappears no matter how small U is.

1. Introduction

A one-dimensional quasiperiodic (QP) tight-binding model is described by the Hamiltonian

$$H_0 = -\sum_{j,s}(c^\dagger_{j+1s}c_{js} + c^\dagger_{js}c_{j+1s}) + \lambda\sum_{j,s} V(j\omega)c^\dagger_{js}c_{js}, \qquad (1)$$

where c_{js} ($c^\dagger_{js}$) is an annihilation (a creation) operator of an electron with spin s at the jth site, $V(x)$ is a periodic function with period 1, i.e., $V(x+1) = V(x)$, and ω is an irrational number.

One of the most typical QP models is the Harper model:

$$V(x) = \cos(2\pi x). \qquad (2)$$

This model has been studied for many years in connection with the problem of two-dimensional Bloch electrons under a magnetic field [1,2]. In the incommensurate Harper model, all the states are extended (absolutely continuous spectrum) for $\lambda < 2$, and are localized (dense point spectrum) for $\lambda > 2$ [3]. At the critical coupling $\lambda = 2$, all the states are "critical" (singular continuous spectrum). Another well-studied model is the Fibonacci model [4,5] defined by $\omega = 1/\tau$ ($\tau \equiv (\sqrt{5}+1)/2$: the "golden mean") and

$$V(x) = \begin{cases} -1, & (m - 1/\tau < x \le m) \\ 1, & (m < x \le m + 1 - 1/\tau) \end{cases}, \qquad (3)$$

where m stands for arbitrary integers. The sequence $\{V(j\omega)\}_{j=1}^{\infty}$ turns out to be the well-known Fibonacci sequence. The Fibonacci model can be considered as a one-dimensional version of quasicrystals. In this model, all the states are critical, irrespective of λ [4,5].

These two models have a remarkable feature: the spectra are pure, i.e., the extended, localized and critical states do not coexist in the spectra. However, this is not always a general feature of the one-dimensional QP systems. The purity of the spectra in the Harper model and the Fibonacci model is probably because of their special forms of $V(x)$: one has only one Fourier component and the other takes only two values. For more general forms of $V(x)$, it is expected that the different types of states coexist in the spectrum.

In section 2, which is based on ref. 6 by Hiramoto and Kohmoto, the one-electron problem in a QP model with

$$V(x) = \tanh[\mu \cos(2\pi x)]/\tanh \mu \tag{4}$$

is studied to demonstrate the appearance of a mixed spectrum. When μ approaches 0, $V(x)$ approaches $\cos(2\pi x)$, i.e., the model reduces to the Harper model. When μ goes to infinity, $V(x)$ takes only two values:

$$V(x) = \begin{cases} -1, & (m - 1/2 < x < m + 1/2) \\ 1, & (m + 1/2 < x < m + 3/2) \end{cases} \qquad (m\text{: integer}).$$

This form is very similar to (3). A numerical calculation shows that the spectrum in this case is singular continuous as in the Fibonacci model. Thus the model (4) is a model interpolating the Harper model and the Fibonacci-like model.

Once we know that the spectrum in the model (4) is not pure (mixed), the next question to be addressed is: what is the structure of the mixed spectrum? One of the possibilities is that the structure of the spectrum would be similar to that of three-dimensional random systems, i.e., that the spectrum would be divided into the localized regime and the extended regime by a finite number of mobility edges. Another candidate for the structure is that there would be infinitely many mobility edges [3]. One-dimensional QP systems have in general a topologically self-similar band structure. Thus we may expect that every "subband" would have mobility edges and there would be an infinite hierarchy of the mobility edges. In section 2, we show that the mixed spectrum has a *finite* number of mobility edges. Furthermore, it is shown that there is no singular continuous component (critical state) in the spectrum, except in the case of $\mu \to 0$ or $\mu \to \infty$. That is, the critical states appear only in extremely ideal cases. It is examined also in another one-dimensional QP model that the mixed spectrum has a finite number of mobility edges and the critical states appear only in very special cases [7]. Thus these properties are considered to be a general feature of one-dimensional QP systems.

In section 3, the effect of an electron-electron interaction on the spectrum and the wavefunctions in the Fibonacci model and the Harper model is investigated within the Hartree-Fock (HF) approximation. We add the on-site electron-electron interaction

$$H_U = U \sum_j c_{j\uparrow}^\dagger c_{j\uparrow} c_{j\downarrow}^\dagger c_{j\downarrow} \tag{5}$$

to eq. (1). When the HF approximation is applied to (5), the effective Hamiltonian is written as

$$H_{HF} = -\sum_{j,s}(c_{j+1s}^\dagger c_{js} + c_{js}^\dagger c_{j+1s}) + \sum_{j,s}(V(j\omega) + W_s(j\omega)) c_{js}^\dagger c_{js}, \tag{6}$$

where

$$W_s(j\omega) = U\langle c_{j-s}^\dagger c_{j-s}\rangle. \tag{7}$$

$W(x)$ is a periodic function with period 1 and is determined selfconsistently. Even if $V(x)$ has the form of (2) or (3), the effective potential $\lambda V(x) + W_s(x)$ is no longer of such special form. From the study on the one-body problem in section 2, it is reasonable to expect that the HF one-body spectrum and the one-body wavefunctions should have different characters from those of the Harper model and the Fibonacci model with $U = 0$. The critical states may disappear. Mobility edges may appear.

The numerical study in section 3 shows that in the Fibonacci model the HF spectrum remains singular continuous for small values of U, while the singular continuous spectrum in the Harper model with the critical coupling $\lambda = 2$ disappears as soon as U is added.

2. One-Body Problem in a One-Dimensional QP System

In the following, we take ω in eq. (1) as $1/\tau = (\sqrt{5} - 1)/2$. The golden mean τ satisfies the relation $1/\tau = \lim_{n\to\infty} F_{n-1}/F_n$, where F_n is the nth Fibonacci number defined by $F_n = F_{n-1} + F_{n-2}$ and $F_0 = F_1 = 1$. Thus the QP systems with $\omega = 1/\tau$ can be described as the $n \to \infty$ limit of the periodic systems with $\omega = F_{n-1}/F_n$. The spectra of the periodic systems consist of F_n bands. Each band is divided into three subbands as n is increased by 2 or 3. Therefore each point in the spectrum for the QP limit ($n \to \infty$) is specified by an infinite sequence of 1, 0, and $\bar{1}$, which represent the upper, middle, and lower subbands, respectively [8].

In order to distinguish whether the state corresponding to the point specified by a sequence is extended, localized or critical, we use two meth-

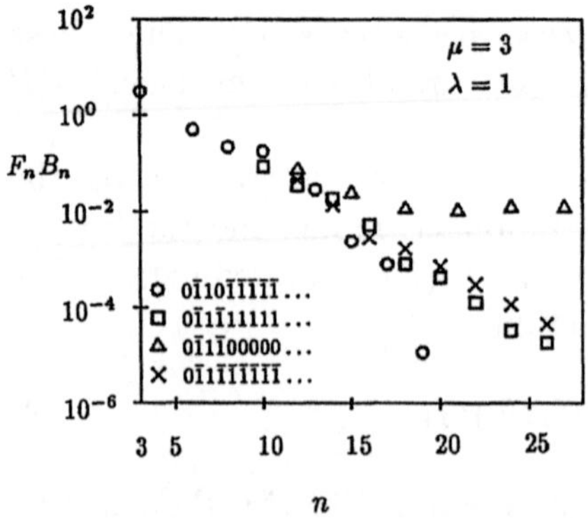

Fig. 1 Plot of $F_n B_n$ vs n for $\mu = 3$ and $\lambda = 1$. The states specified by $\{0\bar{1}1\bar{1}\ldots\}$ are extended, while the state specified by $\{0\bar{1}10\bar{1}\bar{1}\bar{1}\ldots\}$ is localized.

ods. One is the scaling analysis of the bandwidth. By tracing the bandwidth B_n of the periodic system ($\omega = F_{n-1}/F_n$) along the sequence, we investigate the scaling behavior $B_n \sim (1/F_n)^\gamma$ for large F_n. When the state is localized, the bandwidth B_n decreases exponentially as F_n is increased. Thus γ goes to infinity. When the state is extended, γ is 1 for almost all the cases. However there exist extended states with $\gamma = 2$. These states correspond to "band edges" and the behavior comes from a remnant of van Hove singularity in one-dimensional bands. When the state is critical, γ can take an arbitrary value greater than 1.

Figure 1 is a plot of n ($\simeq \log(F_n)/\log\tau$) vs $F_n B_n$ for the states with the code $\{0\bar{1}1\ldots\}$ when $\mu = 3$ and $\lambda = 1$ in the model defined by (1) and (4). From this figure, the values of γ can be estimated. Thus we can determine whether each state is extended, localized or critical. The state specified by $\{0\bar{1}1\bar{1}00000\ldots\}$ ($\gamma = 1$) is extended. The states specified by $\{0\bar{1}1\bar{1}11111\ldots\}$ and $\{0\bar{1}1\bar{1}\bar{1}\bar{1}\bar{1}\bar{1}\bar{1}\ldots\}$ ($\gamma = 2$) are extended as well. On the other hand, the state specified by $\{0\bar{1}10\bar{1}\bar{1}\bar{1}\bar{1}\ldots\}$ is localized.

The other method is the multifractal analysis of a wavefunction $\Psi = \sum_j \psi_j c_j^\dagger |0\rangle$. We study the singular spectrum $f(\alpha)$ [9] of the scaling index α defined by $p_j = |\psi_j|^2 \sim (1/F_n)^{\alpha_j}$ ($\sum_{j=1}^{F_n} p_j = 1$). $f(\alpha)$ is defined as $\Omega(\alpha) \sim (1/F_n)^{f(\alpha)}$, where $\Omega(\alpha)d\alpha$ is the number of lattice sites with α between α and $\alpha + d\alpha$. An extended wavefunction does not have a singular probability measure $p_j \sim 1/F_n$, so $f(\alpha)$ is defined at a single point by $f(\alpha = 1) = 1$. On the other hand, a localized wavefunction has non-vanishing probability only on a finite number of lattice points (measure=0). These points have $\alpha = 0$ and the other lattice points have $\alpha = \infty$. Thus one has $f(\alpha = 0) = 0$ and $f(\alpha = \infty) = 1$. A critical wavefunction with a distribution of α has a smooth $f(\alpha)$ defined on a finite interval $[\alpha_{min}, \alpha_{max}]$. We calculate $f(\alpha)$ numerically in a finite system (a periodic approximation) according to the method of ref. 10. In the analysis, we must be deliberate, because the numerical calculation in a finite

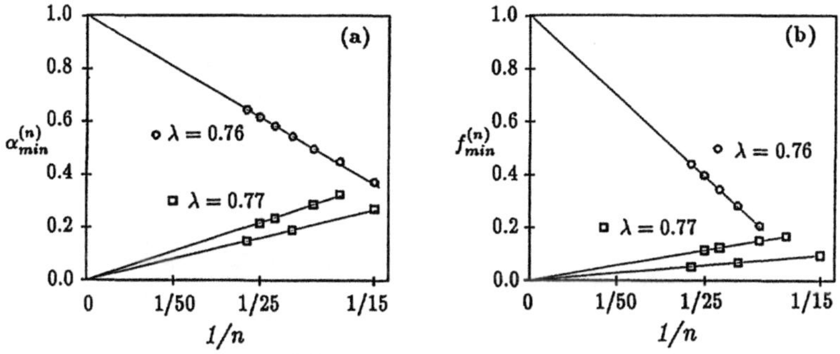

Fig. 2 Plots of (a) $\alpha_{min}^{(n)}$ and (b) $f_{min}^{(n)}$ as a function of $1/n$ at $\lambda = 0.76$ and $\lambda = 0.77$ with $\mu = 2$ for the state $\{\bar{1}\bar{1}\bar{1}\bar{1}\ldots\}$. A sharp transition is observed between $\lambda = 0.76$ and $\lambda = 0.77$.

system always gives a smooth $f(\alpha)$. Therefore a careful extrapolation is required to distinguish extended, localized and critical states.

Figures 2(a) and 2(b) are plots of $\alpha_{min}^{(n)}$ and $f_{min}^{(n)}$ (α_{min} and $f(\alpha_{min})$ calculated for F_n) against $1/n$ for the state $\{\bar{1}\bar{1}\bar{1}\bar{1}\bar{1}\bar{1}\ldots\}$ at $\lambda = 0.76$ and $\lambda = 0.77$ with $\mu = 3$. $\alpha_{min}^{(n)}$ and $f_{min}^{(n)}$ turn out to be linear with respect to $1/n$. It is thus easy to extrapolate to n infinity and estimate α_{min} and $f(\alpha_{min})$. From these figures, it is found that $\alpha_{min} = 1$ and $f(\alpha_{min}) = 1$ for $\lambda = 0.76$, while α_{min}=0 and $f(\alpha_{min}) = 0$ for $\lambda = 0.77$. By a similar analysis, α_{max} and $f(\alpha_{max})$ can be estimated, and the result is that $\alpha_{max} = 1$ and $f(\alpha_{max}) = 1$ for $\lambda = 0.76$, while $\alpha_{max} = \infty$ and $f(\alpha_{max}) = 1$ for $\lambda = 0.78$. Therefore it is concluded that the state $\{\bar{1}\bar{1}\bar{1}\bar{1}\bar{1}\ldots\}$ is extended for $\lambda = 0.76$ and is localized for $\lambda = 0.77$.

The two methods (the bandwidth analysis of the spectrum and the wavefunction analysis) are used to determine whether a state is extended or localized. The results in the remainder of this section are consistently obtained from these two methods.

The spectra of the model defined by (1) and (4) are shown in Fig. 3(a) for $\mu = 1$ and Fig. 3(b) for $\mu = 3$. Only the lower halves of the spectra ($E < 0$) are displayed since the spectra are symmetric with respect to $E = 0$. In region I, the spectra are dense point (localized state), while in region II the spectra are absolutely continuous (extended state). The methods mentioned above are used to distinguish them. One of the most remarkable features we find in these figures is that localization of the wavefunctions does not always start from the edge of the spectrum.

At small values of λ, all the states are extended both for $\mu = 1$ and $\mu = 3$. For $\mu = 1$, localization takes place first at the center of the spectrum as λ is increased; the boundary between the localized and the extended states (mobility edge) moves from the center towards the edge

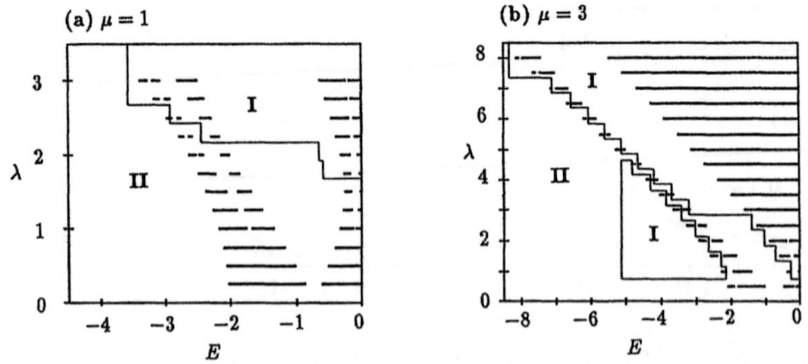

Fig. 3 Energy spectra of the model defined by eqs. (1) and (4) in the text with (a) $\mu = 1$ and (b) $\mu = 3$ for various values of λ. The wavefunctions are localized in region I, and are extended in region II.

of the spectrum with increasing λ; finally all the states become localized at a value of $\lambda(=2.526)$. For $\mu = 3$, a more complicated behavior is found. Localization from the edge of the spectrum occurs in addition to that from the center, and two mobility edges in $E < 0$ (i.e., four mobility edges on the whole spectrum) exist in a range of λ. For larger λ, however, the localized states near the edge of the spectrum disappear. On the other hand, the region of the localized states around the center gradually extends towards the edge, and finally at $\lambda = 7.28$ all the states become localized.

We have also performed a calculation for other values of μ [6]. The behavior of smaller values of μ is similar to the case of $\mu = 1$, and that for larger values of μ is similar to the case of $\mu = 3$. Critical states are found only at $\mu \to 0$ or $\mu \to \infty$.

3. Effect of an Electron-Electron Interaction

In this section, we investigate the effect of the on-site interaction (5) within the Hartree-Fock (HF) approximation [11]. From now on, we neglect spin polarization; i.e., $\langle c_{j\uparrow}^{\dagger} c_{j\uparrow} \rangle = \langle c_{j\downarrow}^{\dagger} c_{j\downarrow} \rangle$ is assumed, because our attention is on whether the mean field destroys the pure spectrum and the singular continuous spectrum. Thus the suffix s standing for spin quantum number is omitted. Although the methods used in the previous section are applied to HF one-body states, only the result of the bandwidth analysis is shown below.

First we study the effect of U on the Fibonacci model. The numerical results shown below are for the Fermi level situated in the first main gap in the spectrum (the gap between $\{\bar{1}11111\ldots\}$ and $\{0\bar{1}\bar{1}\bar{1}\bar{1}\bar{1}\ldots\}$). In

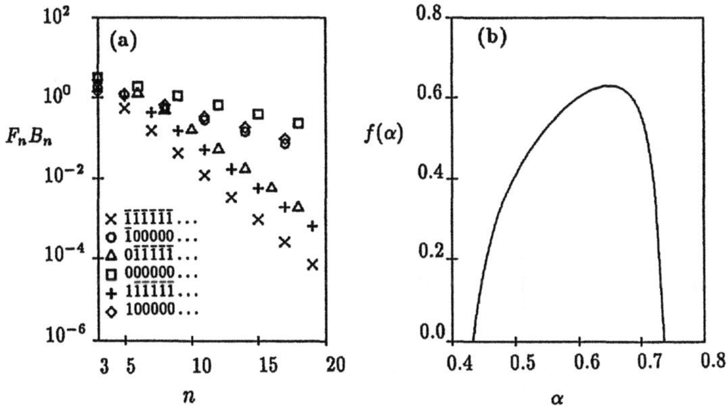

Fig. 4 Scaling analysis for the HF one-body spectrum of the Fibonacci model with the on-site electron-electron interaction ($\lambda = 1$, $U = 0.5$). Plot of F_nB_n vs n is shown in (a), and $f(\alpha)$ for the spectrum is shown in (b).

Fig. 4(a), a plot of n vs F_nB_n for several states is displayed for $\lambda = 1$ and $U = 0.5$. It is clear from the figure that these states remain critical. For example, the value of γ ($B_n \sim (1/F_n)^\gamma$) for the state specified by $\{\bar{1}\bar{1}\bar{1}\bar{1}\ldots\}$ is about 2.32. However, this figure shows only that some special states are critical. To investigate whether all the one-body states are critical or not, we have to study the distribution of γ in the whole spectrum. For this purpose, we again employ the multifractal technique.

We study the distribution of α defined by $1/\gamma$. (This α is different from that used in section 2.) Figure 4(b) shows $f(\alpha)$ for $\lambda = 1$ and $U = 0.5$. This curve is obtained using an extrapolation technique similar to that explained for the wavefunction scaling analysis in the previous section. The extrapolation is carried out using the numerical data up to $n = 20$. The spectrum scaling index α (γ) distributes continuously between $\alpha_{min} \simeq 0.431$ and $\alpha_{max} \simeq 0.74$ ($\gamma_{min} = 1/\alpha_{max} \simeq 1.4$ and $\gamma_{max} = 1/\alpha_{min} \simeq 2.32$). It is confirmed from this result that the HF one-body spectrum is purely singular continuous, i.e., that all the HF wavefunctions are critical.

We have also performed the same analysis for other values of U (including the cases with $U < 0$) and other positions of the Fermi level. The result is as follows. As long as $|U|$ is not so large, the HF spectrum remains purely singular continuous. When $|U|$ is large, many locally stable solutions appear, and it becomes difficult to obtain the true ground state. Thus we cannot offer any reliable comment on the case of large $|U|$. When the Fermi level is situated in a smaller size of gap, the threshold value of $|U|$ at which many locally stable solutions appear becomes smaller.

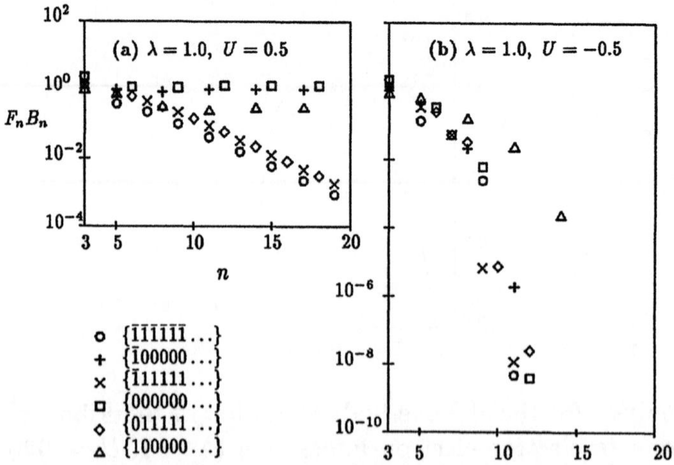

Fig. 5 Plots of n vs F_nB_n for the Harper model ($\lambda = 1$) with (a) $U = 0.5$ and (b) $U = -0.5$.

Next we study the effect of U on the Harper model. The location of the Fermi level in the following numerical results is the same as that above (in the first main gap in the spectrum). The most interesting question is whether the critical states at $\lambda = 2$ are stable for U as the Fibonacci model or not. Thus, here, we only show the result for $\lambda = 2$. In Fig. 5 plots of n vs F_nB_n for several states are shown for (a) $U = 0.5$ and for (b) $U = -0.5$. When $U = 0.5$, all the states shown in this figure turn out to be extended. Since the states specified by $\{\ldots 11111 \ldots\}$ or $\{\ldots \bar{1}\bar{1}\bar{1}\bar{1}\bar{1} \ldots\}$ are identical to band edges, the scaling index γ takes 2 for these states (van Hove singularity). By applying the multifractal analysis to the spectrum, we can conclude that the spectrum is purely absolutely continuous, i.e., that all the states are extended.

When $U = -0.5$, on the other hand, all the states shown in this figure are localized. By applying the multifractal analysis to the spectrum, we can show that the spectrum is purely dense point, i.e., that all the states are localized.

We have systematically investigated the spectrum for various values of U. The result is that the singular continuous spectrum disappears as soon as the electron-electron interaction is added, no matter how small $|U|$ is. However, the purity of the spectrum is not broken for small $|U|$. For $U > 0$, the spectrum is purely absolutely continuous. For $U < 0$, the spectrum is purely dense point.

The phenomenon that the critical states disappear in the Harper model with nonzero U is natural in view of the result of section 2. Further, the fact that the states become extended for $U > 0$ and localized for $U < 0$ can be explained as follows. The mean field $W(x)$ defined by eq. (7) is expanded as $W(x) = \sum_{l=1}^{\infty} W_l \cos(2\pi l x)$. When $U > 0$, the interaction has the effect of screening the one-body potential. Thus W_1 is expected to be negative. If U is small, higher Fourier components W_l $(l \geq 2)$ are small. Thus the effective Hamiltonian (6) turns out to be very close to the Harper model with $\lambda = 2 + W_1 < 2$. This is the reason why all the states are extended when U is positive. The localization for $U < 0$ can be explained in the same manner. When U is not small, the higher Fourier components become important, and the mixed spectrum can appear [7].

On the other hand, of great surprise is that the states in the Fibonacci model with nonzero U remain critical. We can obtain a rather intuitive understanding for the stableness of the critical states in the Fibonacci model by studying the form of $W(x)$. This point as well as the mixed spectrum in the Harper model with large values of U will be discussed elsewhere.

4. Summary

A one-dimensional QP system defined by (1) can have a mixed spectrum which has a *finite* number of mobility edges separating extended regimes and localized regimes. Critical states (singular continuous spectrum) appear only in very exceptional cases including the Harper model with $\lambda = 2$ and the Fibonacci model. We have exemplified these by studying the model (4). In this model, furthermore, the localization starting from the center of the spectrum is found.

The effect of the on-site electron-electron interaction on the Fibonacci model and the Harper model has been also studied within the HF approximation. Critical states in the Harper model with the critical coupling $\lambda = 2$ are unstable against the interaction U, i.e., they disappear if U is not zero. Conversely, the critical states in the Fibonacci model are stable against small U.

References

1. P. G. Harper, Proc. Phys. Soc. London **A68**, 874 (1955).
2. D. R. Hofstadter, Phys. Rev. B **14**, 2239 (1976).
3. S. Aubry and G. André, Ann. Isr. Phys. Soc. **3**, 113 (1980).
4. M. Kohmoto, L. P. Kadanoff and C. Tang, Phys. Rev. Lett. **50**, 1870 (1983).

5. S. Ostlund, R. Pandit, D. Rand, H. J. Schellnhuber and E. D. Siggia, Phys. Rev. Lett. **50**, 1873 (1983).
6. H. Hiramoto and M. Kohmoto, Phys. Rev. Lett. **62**, 2714 (1989).
7. H. Hiramoto and M. Kohmoto, Phys. Rev. B **40**, 8225 (1989).
8. M. Kohmoto, B. Sutherland and C. Tang, Phys. Rev. B **35** 1020 (1987).
9. T. C. Halsely, M. H. Jensen, L. P. Kadanoff, I. Procassia and B. I. Shraiman, Phys. Rev. A **33**, 1141 (1986).
10. T. Janssen and M. Kohmoto, Phys. Rev. B **38** 5811 (1988).
11. H. Hiramoto, J. Phys. Soc. Jpn. **59**, 811 (1990).

Quasiperiodic Systems with Long-Range Hierarchical Interactions

*T. Tokihiro**

Department of Applied Physics, University of Tokyo, Tokyo 113, Japan
*Present address: Department of Physics, University of Pennsylvania, Philadelphia, PA 19104, USA

The electronic and magnetic properties of quasiperiodic systems with long-range hierarchical interactions are investigated. A dynamical system which governs the electronic states completely is found for an electronic system with long-range electron transfer. All the eigenstates are obtained for a quantum Heisenberg model with hierarchical exchange interactions under an external field. The complete self-similarity of the magnetization is proven rigorously for this model.

1. Introduction

Since the discovery of quasicrystals[1,2], their physical properties have atracted much interest [3]. Several ideal quasicrystalline models and relevant quasiperiodic systems have been theoretically investigated. The square-well model, which is a one-dimensional (1-D) quasicrystalline model on a Fibonacci lattice with nearest neighbor interactions, has been studied energetically and its physical properties have been well understood[4,5]. On the other hand, higher dimensional models or long-range interaction models have not been understood sufficiently, though several rigorous results have been obtained for several systems on a 2-D Penrose lattice[6,7]. The reason why the square-well model can be examined successfully is that it can be renormalized exactly and related to a dynamical mapping. Such a simple renormalization transformation has not been found for the other models yet, because the transfer matrix approach can hardly be applicable to them.

In the present paper, we shall study the magnetic properties of a model quasicrystalline system with long-range hierarchical interactions [8,9]. The electronic states of a similar hierarchical model will be briefly summarized in the next section. The present model is defined on a quasiperiodic lattice with the same structure as a 1-D Fibonacci lattice or a 2-D triadic Penrose lattice. Because of their self-similarity, every subcluster can be decomposed into two smaller subclusters. The hierarchical interactions on a 1-D Fibonacci lattice are illustrated in Fig. 1. All interactions between two sites on a subcluster of the same kind are assumed to be equal if both of them do not simultaneously belong to one

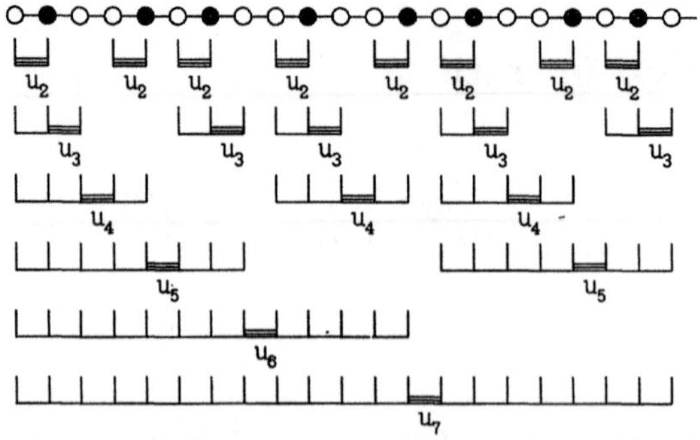

Fig. 1 The set of uniform interaction coefficients $\{u_n\}$ for a 1-D Fibonacci lattice. The interaction between a spin on the right side of the boundary ($\equiv$) and one on the left side is equal in the same kind of subclusters.

of the smaller subclusters. Since the interactions are long-ranged, we cannot use the transfer matrix method, but we have to renormalize the Hamiltonian directly. This can be done rigorously due to the mean-field like nature of the interactions. In fact, the present model may be regarded in a sense as a mean-field approximant of a real quasicrystal. Precisely speaking,it proceeds a bit from mean-field thory, which will be clarified when we discuss its magnetic properties.

2. Long-range electron transfer model

The Hamiltonian of the hierarchical electron transfer model has a simple recurrence form in a matrix representation[8]:

$$\mathbf{H}_n = \begin{pmatrix} \mathbf{H}_{n-1} & \mathbf{U}_n \\ \mathbf{U}_n^t & \mathbf{H}_{n-2} \end{pmatrix}, \tag{2.1}$$

where $\mathbf{H}_n$ is a $F_{n+1} \times F_{n+1}$ matrix for the electronic system on a n-th level subcluster, $\mathbf{U}_n$ is a $F_n \times F_{n-1}$ matrix with all equal matrix elements u_n, $\mathbf{H}_0 = u_0$, $\mathbf{H}_1 = u_1$ and F_n denotes the n-th Fibonacci number, that is,

$$F_{n+1} = F_n + F_{n-1}, \quad F_0 = 0, \quad F_1 = 1. \tag{2.2}$$

If we define $X_n(E)$ as the inverse of the sum of all the elements of the Green function:

$$X_n(E)^{-1} \equiv \sum_{i,j} [(E\mathbf{1} - \mathbf{H}_n)^{-1}]_{i,j}, \tag{2.3}$$

we can find a recurrence formula for $X_n(E)$:

$$X_n(E) = \frac{X_{n-1}(E)X_{n-2}(E) - u_n^2}{X_{n-1}(E) + X_{n-2} + 2u_n}, \tag{2.4}$$

with the initial condition:

$$X_0(E) = E - u_0, \quad X_1(E) = E - u_1. \tag{2.5}$$

The zeros of $X_n(E)$ are equal to the eigenvalues of H_n, and therefore correspond to the eigenenergies of the n-th level subcluster.

The eigenvector $\psi \equiv (\psi_0, \psi_1, \cdots, \psi_{F_{n+1}-1})^t$ of an eigenvalue $E_\lambda^{(n)}$ can be given with respect to $\{X_k(E_\lambda^{(n)})\}_{k=0}^n$ as

$$\psi_m = \prod_{k=2}^{n-1} \phi_k^{(m)}, \tag{2.6}$$

$$\text{where} \qquad \phi_k^{(m)} = \begin{cases} A_k(E_\lambda^{(n)}) & \text{for } c_k(m) = 0, \\ B_k(E_\lambda^{(n)}) & \text{for } c_k(m) = 1, \\ 1 & \text{for } c_k(m) = \tilde{0}. \end{cases}$$

Here $c_k(m)$ is the k-th Fibonacci code of m, *i.e.*, $m = \sum_{k=2}^{\infty} c_k(m) F_k$, $\tilde{0}$ denotes any '0' which immediately precedes a '1',

$$A_k(E) \equiv \frac{X_{k-2} + u_k}{\sin\theta_k (X_{k-1} + X_{k-2} + 2u_k)},$$

$$B_k(E) \equiv \frac{X_{k-1} + u_k}{\cos\theta_k (X_{k-1} + X_{k-2} + 2u_k)},$$

$$\text{and} \quad \sin\theta_k \equiv (F_{k-1}/F_k)^{\frac{1}{2}}.$$

Thus we find that not only the eigenenergies but also the wave functions of the system are completely determined by the recurrence formula (2.4) and its initial condition (2.5). Equations (2.4) and (2.5) can be considered to be a 2-D discrete (nonautonomous) dynamical system if the matrix elements $\{u_n\}$ are regarded as external forces. We can obtain complete information about the electronic structure of the model from this dynamical system.

One remarkable property of the present model is that the wave function has a similar shape on the same kind of subclusters, that is, the amplitude on a subcluster is equal to that on another subcluster of the same kind if it is multiplied by a certain factor. The wave function has a multifractal structure when $\{A_k\}$ and $\{B_k\}$ are periodic in k, or equivalently, $\{X_k\}$ are periodic, *i.e.*, corresponding to a cyclic point.

If we suppose u_n to decrease in power, *i.e.*, $u_n = \alpha^{2-n}$, $\alpha > \tau \equiv \frac{1+\sqrt{5}}{2}$ (golden mean), then the recurrence formula (2.4) comes to have a simple form and we can show the following results explicitly.

(1). Every state is characterized by a quantum number p ($1 \leq p \leq \tau^2$).
(2). The quantum number of a self-similar state is expressed as $p = r\tau + s$, where r and s are some rational numbers.
(3). Almost all states are 'chaotic'.
(4). The energy spectrum is strongly conjectured to be a Cantor set with Lebesgue measure zero.

The detailed proofs and explanation of these statements are given in Ref. [8].

3. Heisenberg model with hierarchical interactions

If we regard $\{u_n\}$ as exchange interactions $\{J_n\}$ between spins, we get a spin system with quasiperiodic long-range interactions[9]. We can treat any kind of spin model, but we restrict ourselves to spin $\frac{1}{2}$ quantum Heisenberg model in the present paper.

The Hamiltonian of the present model can be expressed as

$$H_N = \sum_{p=1}^{N} J_p \sum_{r,r'} \mathbf{S}_{p,r} \cdot \mathbf{S}_{p+1,r'} - HS^z_{N+2}. \tag{3.1}$$

Here $\mathbf{S}_{p,r}(\mathbf{S}_{p+1,r'})$ is the spin-sum of the r-th (r'-th) block of p-th($(p+1)$-th) subcluster which contains $F_p(F_{p+1})$ spins, *i.e.*, $\mathbf{S}_{p,r} = \sigma_{n_p^r} + \sigma_{n_p^r+1} + \cdots + \sigma_{n_p^r+F_p-1}$, where σ_j denotes a spin operator on a site j, and $n_p^r = (r-1)F_p + (r'-1)F_{p-1}$. The second summation of the right hand side of eq.(3.1) should be taken over all the adjacent pairs of $\mathbf{S}_{p,r}$ and $\mathbf{S}_{p+1,r'}$ which form $(p+2)$-th block spins. This model has a similar form to Dyson's hierarchical model[10-13]. From the viewpoint of renormalization group, Baker showed that this assumption for interactions is equivalent to Wilson's approximate renormalization transformation, which is exact to first order in $\epsilon = 4 - d$ [11,14]. Therefore, the present model may be regarded as a first order correction to the mean-field theory. The crucial disadvantage of Dyson's hierarchical model, when applied to a usual crystal, is that interactions are not translationally invariant. For a quasicrystal, however, this feature is not a disadvantage at all, since it has no translational symmetry in itself.

It is useful to rewrite the Hamiltonian (3.1) as

$$\tilde{H}_N = \sum_{p=1}^{N} \sum_{r=0}^{F_{N+1-p}-1} \tilde{J}_{p,r}(\mathbf{S}_{p+2,r})^2 - HS^z_{N+2}, \tag{3.2}$$

where $\tilde{J}_{p,r}$ form a Fibonacci sequence with $\tilde{J}_p \equiv \frac{1}{2}(J_p - J_{p+1})$ and $\tilde{J}'_p \equiv \frac{1}{2}(J_p - J_{p+2})$, that is, $\{\tilde{J}\}_{r=0}^{F_{N+1-p}-1} = \{\tilde{J}_p, \tilde{J}'_p, \tilde{J}_p, \tilde{J}_p, \tilde{J}'_p, \tilde{J}_p, \tilde{J}'_p, \tilde{J}_p, \cdots\}$ and we have omitted unimportant constant terms. Then, we can easily find that there is one and only one state of $\tilde{H}_N$ having the eigenvalues $j_{p,r}(j_{p,r}+1)$ for $(\mathbf{S}_{p,r})^2$ and m for S^z_{N+2} with eigenenergy

$$E_N(\{j_{p,r}\}, m; H) = \sum_{p=3}^{N}\sum_{r} \tilde{J}_{p,r} j_{p,r}(j_{p,r}+1) - Hm. \tag{3.3}$$

Thus we obtain all the eigenstates for the present model in spite of the fact that it is usually rather difficult to find only a ground state for a quantum Heisenberg model even in the absence of an external field. The above results do not depend on whether interactions are antiferromagnetic or not, but hereafter we restrict ourselves to the antiferromagnetic interactions satisfying the inequalities $J_p > J_{p+1} > 0$ and $\lim_{p\to\infty} J_p = 0$.

The ground state without an external field ($H = 0$) is clearly given by minimizing all the quantum numbers $\{j_{p,r}\}$. The ground-state wave function for $N = 3n-2$ is expressed as

$$\begin{aligned} |\psi> = [1,2][4,5][6,7][9,10]\cdots[\frac{F_{3l-1}+1}{2}, \frac{F_{3l-1}+1}{2}+F_{3l-1}] \\ [\frac{F_{3l-1}+1}{2}+F_{3l}, \frac{F_{3l-1}+1}{2}+F_{3l-1}+F_{3l}] \\ \cdots[\frac{F_{3n-1}+1}{2}, \frac{F_{3n-1}+1}{2}+F_{3n-1}], \end{aligned} \tag{3.4}$$

where $[i,j]$ stands for a normalized singlet combination of spin i and j. (We have only to add one free spin for $N \neq 3n-2$.)

The present system is a frustrated quantum spin system and is not expected to have true long-range order without an external field. In fact, if we define a two point correlation function

$$g(r) \equiv \lim_{N\to\infty} \frac{1}{(F_{N+2}-r)} \sum_{i=1}^{F_{N+2}-r} < \sigma_i \cdot \sigma_{i+r} >, \tag{3.5}$$

we find self-similar quasi-long-range antiferromagnetic order:

$$g(r) = \begin{cases} (-\frac{3}{2})(\frac{1}{\tau})^{3k-1} & \text{for} \quad r = F_{3k-1}, \\ 0 & \text{otherwise.} \end{cases} \tag{3.6}$$

It should be noticed that the energy spectrum has no gap immediately above the ground state since the lowest excitation energy is equal to $2J_N$ which goes to 0 in the limit of infinite volume. (Dyson's hierarchical model has a gap contrary to the present model.)

Now we consider the ground state in the presence of a uniform external field ($H > 0$). Since an eigenenergy $E_N(\{j_{p,r}\}, m; H)$ is given by

$$E_N(\{j_{p,r}\},m;H) = E_N(\{j_{p,r}\},m;0) - mH, \tag{3.7}$$

we should first seek the lowest energy state with a given magnetization m in order to determine the ground state. If we define $E_N(m) \equiv \min_{\{j_{p,r}\}}[E_N(\{j_{p,r}\},m;0)]$, we find

$$E_N(m) = \min_{0\le m'\le m}[E_{N-1}(m-m') + E_{N-2}(m') + \tilde{J}_{N-1}m'(m'+1)] \\ +\tilde{J}_N m(m+1). \tag{3.8}$$

From eq.(3.8), we can also verify the following statements by induction:
(a)

$$\Delta E_N(m+1) > \Delta E_N(m) > 0 \tag{3.9}$$

where $\Delta E_N(m) \equiv E_N(m) - E_N(m-1)$,
(b) if $m' = m_0$ is the number that minimizes the right hand side of eq.(3.8) for $E_N(m)$, then $m' = m_0$ or $m' = m_0 + 1$ for $E_N(m+1)$.

From eq.(3.8) and (b), we can deduce that the wave function corresponding to $E_N(m)$ consists of $(\frac{1}{2}F_N - m)$ singlet pairs and $[m]$ triplet pairs, where $[m]$ denotes the integral part of m, and all the spin pairs are equal to those without an external field. Since the magnetization is m, these pairs uniquely determine the lowest energy state with $j_{N+2} = m$ and $S_{N+2}{}^z = m$, although it depends on the concrete values of coefficients $\{J_p\}$ which pair is singlet or triplet.

Furthermore, from eqs.(3.7) and (3.9), the ground state under a given external field H can be determined as the lowest energy state with magnetization $m(H) = m$ which satisfies the inequality

$$\Delta E_N(m) < H < \Delta E_N(m+1). \tag{3.10}$$

As an example, let us assume J_p to decrease in power : $J_p = \alpha^{1-p}$. If α is large enough, that is, $\alpha \ge \tau^2$, we can obtain the ground state explicitly. Since the frustration is weak, the rule for constructing the ground state is simply that the partial magnetization of a subcluster should be smaller than those of all the larger subclusters. Using the features of the Fibonacci code, we can prove the following results:

(A) Every pair with a radius (a distance between the two spins) greater than $r_c \equiv F_{3l-1}$ is triplet, where l is the integer which is defined for a given magnetization m by an inequality : $\frac{1}{2}F_{3(n-l)} < m \le \frac{1}{2}F_{3(n-l+1)}$.

(B) Every pair with a radius less than r_c is singlet.

(C) If we define an index i_k for a k-th pair with a radius r_c ($[n_k, n_k+r_c]$) as $i_k \equiv \sum_{j=2}^{3(n-l)} c_j(k)\ F_{3(n-l)-j}$, then the pair is singlet if and only if $i_k \leq \frac{1}{2}F_{3(n-l+1)-m}$. Here, $n_k = n_0 + F_{3l+k}$ and $n_0 \equiv \frac{F_{3l-1}+1}{2}$.

As is known from (A)-(C), the correlation function $g(r)$ of the ground state under uniform external field satisfies

$$\begin{cases} g(r) \leq 0 & \text{for} \quad r < r_c, \\ g(r) > 0 & \text{for} \quad r > r_c. \end{cases} \tag{3.11}$$

Namely, the ground state has short-range antiferromagnetic order and long-range ferromagnetic order. The critical length r_c is determined by the field strength as is stated in (A). This is one of the peculiar features of the present system and a similar behavior is conjectured to be observed in the frustrated Ising model on a 2-D Penrose lattice[15].

Under the same assumption on the coefficients $\{J_p\}$, the normalized magnetization $M(H) \equiv \lim_{N\to\infty} m(H)/S_{N,max}$ exhibits complete self-similarity

$$M(\frac{H}{\alpha^3}) = \frac{1}{\tau^3}M(H), \tag{3.12}$$

and devil's staircase structure as is shown in Fig.2. Here, $M^{-1}(0) = 0$ and $M^{-1}(1) = \sum_{k=1}^{\infty} \sim \frac{\alpha-1}{\alpha-\tau} \leq \tau$. The proof is straightforward from eq.(3.10), because the exact expression of $\Delta E_N(m)$ can be obtained after a little painful algebra as

$$\Delta E_N(m) = \sum_{k=1}^{3(n-l)} \tilde{J}_{N-k} e_k(s) \tag{3.13}$$

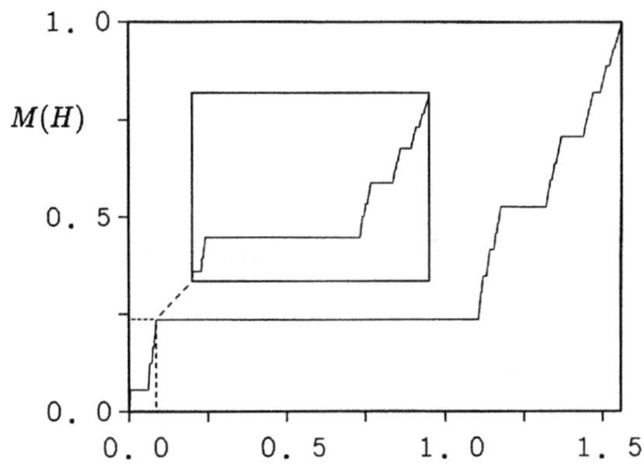

Fig. 2 The self-similar magnetization of $M(H)$ is shown for $\alpha = \tau^2$. The area in the square shows the self-similarity given by eq. (3.12).

where $s \equiv \frac{1}{2}F_{3(n-l+1)} - m$,

$$e_k(s) = \begin{cases} F_{3(n-l)+3-k} - 2 \sum_{j=1}^{3(n-l)+1-k} c_{j+k-1}(s)F_j & \text{if } c_k(s) = 0, \\ F_{3(n-l)+2-k} - 2 \sum_{j=1}^{3(n-l)+1-k} c_{j+k-1}(s)F_{j-1} & \text{if } c_k(s) = 1, \end{cases}$$

$c_k(s)$ is the k-th Fibonacci code of s and $c_1(s)$ is set to be 0. In fact, if we define m^* and s^* for an arbitrary m ($m = \frac{1}{2}F_{3(n-l+1)} - s, 0 \leq m \leq \frac{1}{2}F_{3(n-1)}$) as $m^* \equiv \frac{1}{2}F_{3(n-l+2)} - s^*$ and $s^* \equiv \sum_{j=2}^{3(n-l+1)} c_j(s)F_{j=3}$, then we find

$$\Delta E_N(m^*) = \alpha^3 \Delta E_N(m) + O(F_{3(n-l+1)}J_N), \tag{3.14}$$

$$\text{and} \qquad m^* = \tau^3 m + O(N). \tag{3.15}$$

Equations (3.14) and (3.15) lead to the self-similarity (3.12).

If the frustration is strong, *i.e.* $\alpha < \tau^2$, the magnetization no longer has complete devil's staircase structure, but seems to have an incomplete devil's staircase structure.

Finally, we briefly mention the thermodynamic properties of the present model and other models with similar quasiperiodic hierarchical interactions. The present antiferromagnetic Heisenberg model is a frustrated quantum spin system and has no true long-range order at $T = 0$. Therefore it has no phase transition at finite temperature. The specific heat, however, has several structures due to the energy gaps at the energies of triplet pair excitations. Since the present model can be renormalized exactly, we can obtain the partition function Z_N by the recurrence formu1a

$$Z_N = \sum_{m=S_N,min}^{S_N,max} \frac{\sinh(\frac{1}{2}(2m+1)\beta H)}{\sinh(\frac{1}{2}\beta H)} X_N(m), \tag{3.16}$$

$$X_N(m) = e^{-\beta \tilde{J}_N m(m+1)} \times$$

$$\sum_{|m_1 - m_2| \leq m \leq m_1 + m_2} e^{\beta \tilde{J}_{N-1} m_2(m_2+1)} X_{N-1}(m_1) X_{N-2}(m_2),$$

$$X_0(\frac{1}{2}) = e^{\frac{3}{4}\beta \tilde{J}_1}, X_1(0) = 1, X_1(1) = e^{-2\beta \tilde{J}_1}, \tag{3.17}$$

where $\beta \equiv (k_B T)^{-1}$ as usual. The numerical calculation shows that the susceptibility is a monotonically decreasing function of the temperature, while the specific heat has several structures as is expected[9].

If the exchange interactions are ferromagnetic, on the other hand, we can prove that not only the present quantum Heisenberg model but also the classical Heisenberg and Ising models show phase transitions under similar conditions to those in Dyson's hierarchical model. Especially in the Ising model, we can find a 'Gaussian' fixed point of the nonlinear recursive operator for the partition function, and obtain the critical exponents following Bleher and Sinai[12]. The results, however, seem to be essentially the same as those of Dyson's hierarchical model. We can also treat the spherical model with similar long-range interactions. In this case, we get a dynamical system which determines the zeros of the partition function just as the electronic model discussed in the previous section. However, the analysis would be rather complicated.

4. Concluding remarks

We have considered a model quasicrystalline system with long-range hierarchical interactions. In both electronic system and Heisenberg spin system, the Hamiltonian can be renormalized exactly due to the mean-field like nature of interactions. We have found a new quasiperiodic dynamical system which governs the electronic states completely for the electronic system and obtained all the eigenstates for the quantum Heisenberg system under a uniform external field. Complete self-similarity of the magnetization has been proven rigorously. The other systems with different kind of self-similarities such as a 3-D tetrahedral Penrose lattice can be easily treated in a similar way. Our approach would be effective (and limited) to the same extent as the mean-field approximation in discussing the physical properties of quasicrystals.

References

1. D. Schechtman, I. Blech, D. Gratias, and J. W. Cahn, Phys. Rev. Lett. **53**, 1951 (1984).
2. D. Levine and P. J. Steinhardt, Phys.Rev. Lett. **53**, 2477 (1984).
3. See, for example, *The Physics of Quasicrystals* (World Scientific, Singapore, 1987) ed. by P. J. Steinhardt and S. Ostlund, Chap. 7.
4. M. Kohmoto, L. P. Kadanoff, and C. Tang, Phys. Rev. Lett. **50**, 1870 (1983).
5. S. Ostlund, R. Pandit, D. Rand, H. J. Schellnhuber, and E. Siggia, Phys. Rev. Lett. **50**, 1873 (1983).
6. V. E. Korepin, Phys. Lett. **A118**, 285 (1986); Commun. Math. Phys. **110**, 157 (1987).
7. T. Fujiwara, M. Arai, T. Tokihiro, and M. Kohmoto, Phys. Rev. **B 37** 2797 (1988); M. Arai, T. Tokihiro, T. Fujiwara, and M. Kohmoto,

Phys. Rev. **B 38** 1621 (1988); T. Tokihiro, T. Fujiwara, and M. Arai, *ibid*, 5981 (1988).
8. T. Tokihiro, Phys. Rev. **B40**, 2889 (1989).
9. T. Tokihiro, to be published.
10. F. J. Dyson, Commun. Math. Phys. **12**, 91 (1969); *ibid.* **21**, 269 (1971).
11. G. A. Baker Jr., Phys. Rev. **B5**, 2622 (1972).
12. P. Bleher and J. Sinai, Commun. Math. Phys. **33**, 23 (1973); *ibid.* **45**, 247 (1975).
13. G. Gallavotti and H. Knops, Revista del Nuovo Cimento, **5**, 341 (1975).
14. K. G. Wilson, Phys. Rev. **B4**, 3184 (1971); K. G. Wilson and M. E. Fisher, Phys. Rev. Lett. **28**, 540 (1972).
15. Y. Okabe, private communication and the article in this volume.

Conductance of a Penrose Lattice

H. Tsunetsugu

Institute for Solid State Physics, University of Tokyo,
Roppongi 7-22-1, Minato-ku, Tokyo 106, Japan

Abstract. Conductance of a Penrose lattice is studied within the tight-binding approximation. We calculated the conductance at zero temperature as a function of the Fermi energy and the length of the lattice by using the multi-channel Landauer formula. In particular, the length dependence can be studied for large systems (up to ~1.3 million sites) due to the Lee-Fisher method. The conductance shows power-law decay with the lattice length, at least at the low energies where the conductance-energy curve is relatively smooth. This behavior is consistent with our previous conclusion that most of the wavefunctions in a Penrose lattice are neither extended freely nor localized exponentially.

1. Introduction

The aim of the present work is to investigate electronic structures and transport properties of quasicrystals (QC), especially a Penrose lattice (PL), through the calculation of the conductance and to clarify how the structure affects the transport properties of QC. We have studied the dependence of the conductance on the Fermi energy and on the lattice length [1-3]. In this article the emphasis is mainly placed on the length dependence.

The most peculiar and intriguing property of QC is the structure. The diffraction pattern consists of a set of sharp spots with a rotational symmetry forbidden in conventional crystallography (*e.g.* icosahedral [4], decagonal [5]). This means QC do not have translational symmetry but still possess a kind of long-range order [6].

There arises an essential question how the unusual structure affects physical properties. Conductance is one of the most sensitive quantities. It is well-known that the quantum interference effect leads wavefunctions scattered by disorder to localization under some circumstances. This phenomenon is known as the Anderson localization [7]. As it is a site off-diagonal quantity, the conductance is sensitive to phase perturbation of the wavefunctions by the disorder as well as their amplitudes. Therefore the conductance has been used as a measure of the Anderson localization.

Quasicrystals do not have periodicity, however, the structures are *not random* but *quasiperiodic*. Considering this point, wavefunctions in QC should have characters intermediate between crystals and disordered materials. Another difference is that the structure is *topologically* "disordered" in QC, while the on-site potential or the transfer energy is distributed randomly in the regular lattice in the studies of the Anderson localization. (Here the "disorder" refers only to the lack of periodicity.)

The reason for studying the Penrose lattice among quasiperiodic models is that it is the lowest-dimensional model with topological "disorder". According to the renormalization group (RG) theory, the effect of disorder is more serious in the lower dimensional systems [7,8]. Much study has been given to one-dimensional quasiperiodic models, par-

Editors: T. Fujiwara · T. Ogawa

ticularly the Fibonacci chain and some exotic results were obtained [9-11]: *e.g.* singular continuous density of states (DOS) and critical (self-similar/chaotic) wavefunctions. But the topology in one dimension is trivial. Therefore the electron system on the Penrose lattice, a 2D system, is the most sensitive to the quasiperiodicity of topology.

In the studies of the electronic structure of the Fibonacci chain, a simple but exact RG equation of the 2×2 transfer matrix plays an essential role [9]. But in higher dimensions, it is hopeless to obtain such a RG equation with finite degrees of freedom. Moreover, as the localization of wavefunctions is a delicate problem, the validity and applicability of conventional analytical approximations are not clear in quasiperiodic models. So numerical simulations are useful for the investigation of physical properties of QC.

The electronic structure of the Penrose lattice is not understood so well [1,12]. What we know rigorously is limited to some wavefunctions at special energies. The most interesting are the wavefunctions that are confined within a finite section, so-called confined states [12-15]. The degeneracy of the states is thermodynamically infinite, which is the result of a structure property known as Conway's theorem [16,17]. However, the existence of the confined states depends on the Hamiltonian and the high degeneracy implies that they are unstable to additional perturbation. Therefore, we investigated all the eigenstates numerically in a previous work [1].

Our previous conclusions on the electronic structure of the PL are as follows. By analyzing the level statistics of the eigenstates, we found that the DOS is not smooth but the total band width remains finite, different from the case of the Fibonacci chain. Next, we calculated the 8-norm of the wavefunctions, which is a generalization of the participation ratio. From this, we concluded that most of the wavefunctions show power-law decay. Strictly speaking, the distribution of probability density is like the case of power-law decay but it does not necessarily imply that the wavefunctions do decay in a power-law *in the space.*

2. Lattice and Model

The model studied here is a tight-binding Hamiltonian where an s-like atomic orbital is placed in the *center* of each rhombus in the Penrose lattice. Transfer energies are finite only for nearest-neighbor rhombus pairs, -1, otherwise 0. Analysis of the ring statistics shows odd-member rings form the majority (87.5%) [18]. In our model, every atom has 4 nearest-neighbors, so eigenenergies are restricted within the range $-4 \leq E \leq 4$. But due to the odd-member rings, the DOS is asymmetric and antibonding states are frustrated, which results in the drop in the band top from 4.0 to $\sim$2.68 [1].

We must choose carefully the boundary condition in numerical calculations, because the wavefunctions and the conductance are sensitive to the surface scattering. Presumably the best choice is the periodic boundary condition. But because Penrose lattices are not periodic, some modification is necessary in the lattice. In the previous works on the electronic structure [1] and the Fermi energy dependence of the conductance [2], we used a sequence of periodic approximations of a Penrose lattice [19]. They are approximated in 2 directions and we named them "periodic Penrose lattices (PPL)". But for studying the length dependence, a more suitable lattice is the one that is periodic in the lateral direction of the electric current but has quasiperiodicity in the current direction [3]. This lattice can be obtained by approximating the Penrose lattice in only one direction. Hereafter we will distinguish them from the PPL by the name "semi-periodic Penrose lattices (SPPL)".

Figure 1 shows a unit cell of a PPL and a part of a SPPL. When the ratio of successive Fibonacci numbers is used as an approximant of the golden mean $\tau \sim F_{n+1}/F_n$, the

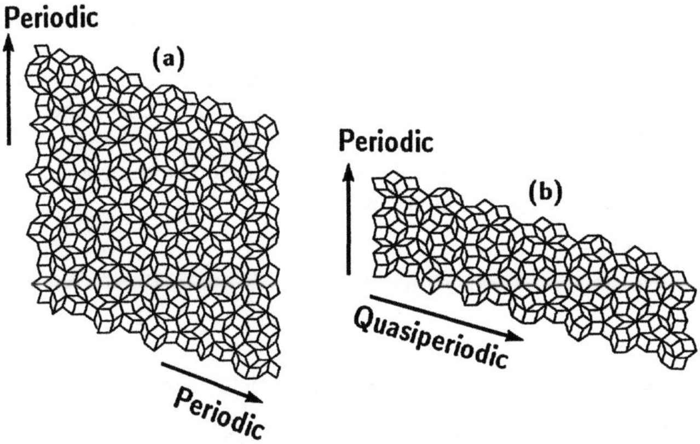

Fig.1 (a) Unit cell of a PPL where τ is approximated by $\frac{8}{5}$. The lattice width is $M = 26$, the site number is $N = 521$. (b) A part of a SPPL where τ is approximated by $\frac{3}{2}$ in the periodic direction. $M = 10$.

lattice approximates the Penrose lattice optimally. In Fig.1, τ is approximated by $\frac{8}{5}$ in (a) and by $\frac{3}{2}$ in (b). Let us define a lattice width M of a PPL or SPPL by the number of rhombuses lining up along the translation vector. For example, M =26 and 10 in Figs.1(a) and (b) respectively. As the Fibonacci generation increases by 1, the lattice width multiplies roughly by τ. It is noted that a unit cell of a PPL contains only two defects (places where the matching rule is violated) [20].

We will confine ourselves to the case $T = 0\,\mathrm{K}$ in this article in order to exclude the effect of thermal fluctuations, which destroy the quantum interference of the wavefunctions and may smear out characteristic behaviors.

3. Method of Calculation

We will use the multi-channel Landauer (MCL) formula [21] as the definition of the conductance. This is convenient for finite systems both in the formulation and in numerical calculations.

The MCL formula is formulated for the system consisting of the relevant scatterer with finite cross section and two semi-infinite lead lines attached to both sides (Fig.2). In our case, the cross section is the lattice width M and the lead lines are square lattices with the same width M. The conductance is defined with the transmission coefficients of the incoming plane waves propagating in the lead line with the same energy as the Fermi energy of the system. These plane waves can be labelled by the momentum in the lateral direction and called channels. It is noted that the number of channels M_0 varies with the Fermi energy and $M_0 \leq M$. The conductance in units of e^2/h per spin is

$$g = \mathrm{Tr}\left(\mathbf{t}^{\dagger}\mathbf{t}\right) = \sum_{\alpha\beta} |t_{\alpha\beta}|^2 , \tag{1}$$

where $\mathbf{t}$ is the transmission matrix and $t_{\alpha\beta}$ denotes the transmission coefficient from β- to α-channel. The law of current conservation implies $g \leq M_0$.

In a previous work [2], we obtained the transmission matrix by solving directly Schrödinger equations with the scattering boundary condition, and we applied the formula (1). But an equivalent expression given by Lee and Fisher [22] is more convenient for the study of the length dependence. The expression is defined with the site Green

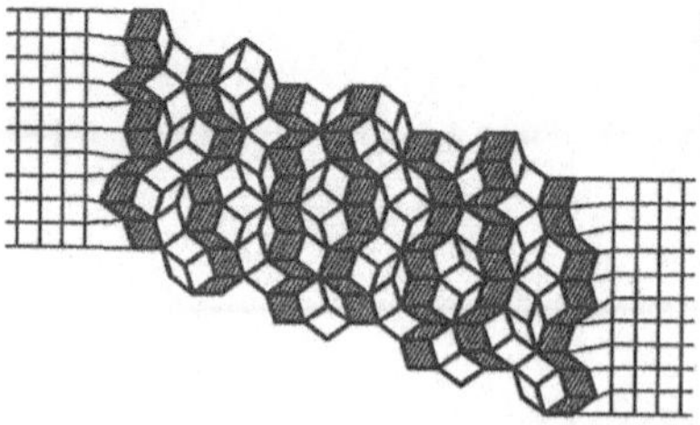

Fig.2 Configuration for the MCL formula. The scatterer, a part of a SPPL, is shown with bold lines and left and right lead lines are attached to both sides. The hatching illustrates the process of extending the scatterer. In the vertical direction, the periodic boundary condition is imposed.

functions at the Fermi energy including an infinitesimal adiabatic constant $E - i0$:

$$g = 4\,\mathrm{Tr}\left[G''(j,j)G''(j+1,j+1) - G''(j,j+1)^2\right], \tag{2}$$

where $G(j,j')$ is the submatrix of the Green function between the columns j and j', the superscript $''$ means the imaginary part, and the trace should be taken over the sites inside the column. The column index j can be arbitrarily chosen due to the current conservation, so we will choose here the far left column in the right lead line for convenience.

The advantage of the Lee-Fisher expression lies in an effective recursion formula of the site Green functions. For this purpose, let us introduce another Green function $\hat{G}^{\mathrm{L}}(l)$. This is the submatrix inside the far right column l in the system which consists of the left lead line and the scatterer up to the l-th column but does not have the right lead line. The Dyson equation relates the Green function to the preceding step where the scatterer contains only up to the $(l-1)$-th column:

$$\hat{G}^{\mathrm{L}}(l) = \left[E - H_l - {}^{t}B_{l-1}\hat{G}^{\mathrm{L}}(l-1)B_{l-1}\right]^{-1}, \tag{3}$$

where H_l is the transfer energy matrix inside the column l, and B_{l-1} is between the columns l and $l-1$. The recursion formula (3) has been modified from the original because of two differences with periodic lattices: one is changes in the number of sites from column to column, and the other is the irregularity of site connectivity. The first point states that the dimension of $\hat{G}^{\mathrm{L}}(l)$ varies with l, and the second point demands the matrix B_{l-1}. The starting point of the recursion is the Green function of the far right column in the left lead line $\hat{G}^{\mathrm{L}}(0) = G_{\mathrm{S}}$. Because the lead line is a square lattice, the analytic form of G_{S} is easily obtained.

The Green functions of the full system containing the right lead line can be calculated with $\hat{G}^{\mathrm{L}}$:

$$\begin{aligned} G(l,l) &= \left[E - H_l - G_{\mathrm{S}} - {}^{t}B_{l-1}\hat{G}^{\mathrm{L}}(l-1)B_{l-1}\right]^{-1}, \\ G(l+1,l+1) &= \left[E - H_{l+1} - G_{\mathrm{S}} - \hat{G}^{\mathrm{L}}(l)\right]^{-1}, \\ G(l,l+1) &= -G(l,l)G_{\mathrm{S}}. \end{aligned} \tag{4}$$

Inserting these Green functions into the formula (2), we get the conductance. It is noted that these Green functions $\hat{G}^{\mathrm{L}}$ and G have the dimension of the number of sites in each column, not the total sites in the scatterer, so the recursion formula works effectively for the study of length dependence.

4. Results and Discussions

First, we show the Fermi energy dependence of the conductance in Fig.3 for two energy regions, as a guide to choosing the Fermi energy in the study of the length dependence. The scatterer is a unit cell of a PPL with 3571 sites, where τ is approximated by $\frac{21}{13}$. The energy resolution is $\frac{1}{5000}$, which is 50 times as fine as the previous result. Figures 3(a) and (b) show the result in a low and high energy regions respectively. As studied in detail in Ref.[2], the smoothness of the $g(E)$ curve changes with the energy region. The curve is relatively smooth at low energies, but spiky at high energies. This behavior agrees with the smoothness of the DOS and the energy dependence of the Thouless number. Of course, whether the energy is high or low depends on the sign of the transfer energies and it does not have a definite meaning. It is noted that the energy scale is different between Figs.3(a) and (b).

We must choose the Fermi energy carefully. When studying intrinsic properties of the length dependence, we must set the Fermi energy within the energy spectrum, because the conductance shows an exponential decay otherwise. Therefore we should avoid the energies where the $g(E)$ curve shows a large dip, which corresponds to an energy gap.

Figure 4(a) shows the length dependence of the conductance at a low energy $E = -3.40$. The scatterers are parts of SPPL with increasing widths. The length of a scatterer is defined as the ratio of the site number to the width, $L \equiv N/M$. We have calculated for 6 systems: the width is from 68 to 754, and the length is up to $\sim$1800.

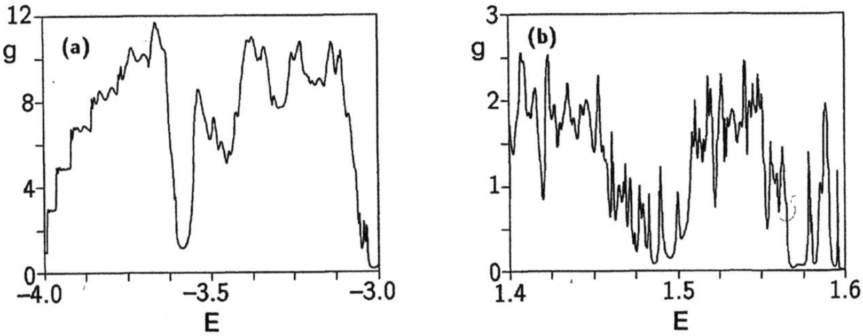

Fig.3 The Fermi energy dependence of the conductance of a PPL in (a) a low energy region, and (b) a high energy region. $M = 68$, $N = 3571$. The energy resolution is $\frac{1}{5000}$.

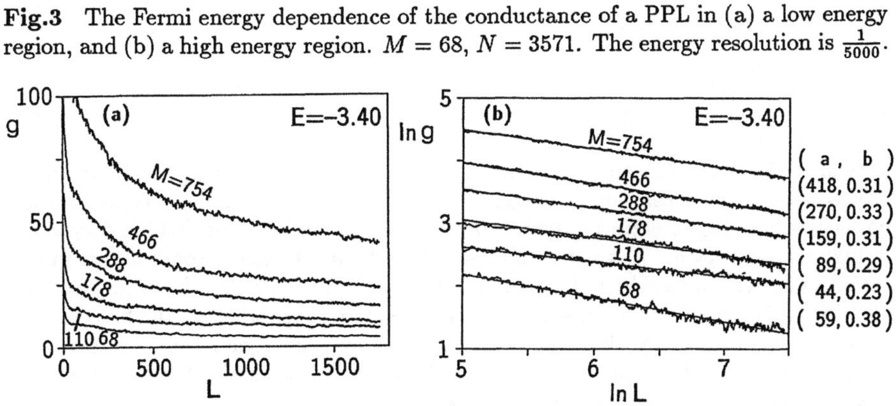

Fig.4 (a) The lattice length dependence of the conductance of 6 SPPL at $E = -3.40$. The lattice widths M are shown in the figure. (b) The log-log plot of (a). Straight lines are power-law functions aL^{-b} fitted by the least-squares method.

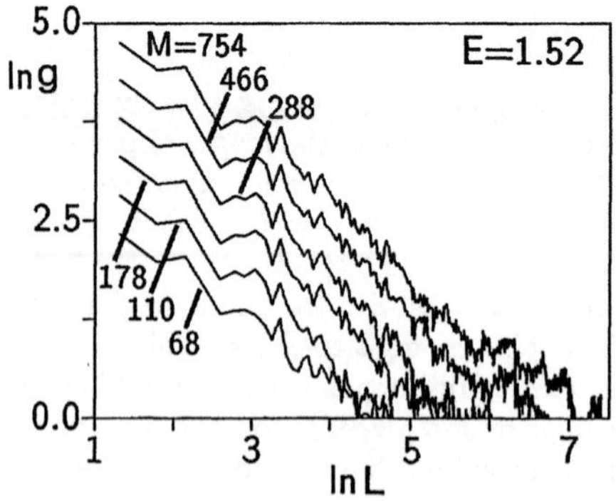

Fig.5 The log-log plot of the length dependence of the conductance at $E = 1.52$. The same lattices are used as Fig.4.

The largest scatterer contains 1 318 368 sites. We can see easily that the conductance decreases slowly with the length of the scatterer. Actually, the log-log plot illustrates power-law decays clearly (Fig.4(b)). The straight lines in Fig.4(b) are the power-law functions fitted by the least-squares method, $g(L) \sim aL^{-b}$. Except for short-range lengths, the behavior of the length dependence can be regarded as a power-law function and fluctuations around it. The exponent of power b varies with the lattice width M and it has not converged yet. But all of them are around 0.3. The variance of the exponent becomes smaller and the linearity in the log-log plot becomes better with M. Therefore it is expected that the exponent will converge in the thermodynamic limit $M \to \infty$.

The power-law decay at the energy $E = -3.40$ was also confirmed by calculations using other parts of the same SPPL. The result shows only small sample dependence, which means the power-law decay is not an effect of the defects introduced into the SPPL by the approximation of the lattice. Moreover, calculations at another energy $E = -3.50$ reproduce the power-law decay [3], although the fluctuations are larger and the exponent b has a smaller value. From these results, the power-law decay is seen to be an intrinsic property at least at low energies.

The power-law decay of the conductance supports our previous conclusion that most of the wavefunctions decay in a power-law.

We make a comment on an exceptional case of near the band bottom where the $g(E)$ curve shows a step-like behavior in Fig.3(a). The height of each step coincides with the number of channels M_0. In fact, the result of the length dependence shows that the conductance does not decay with the length but only fluctuates around M_0. This suggests that the wavefunctions in this energy region are like plane waves, *i.e.* extended states, which agrees with the character of the wavefunctions obtained by the direct diagonalization [23].

The study of the length dependence at high energies involves a greater difficulty. The conductance at some high energies (*e.g.* $E = 1.50$) shows a decay faster than power-law and large fluctuations with L [3]. However, because the DOS is very spiky in high energy regions, it is not guaranteed that the fixed energy would not go out of the energy spectrum as the lattice is being extended. Therefore the fast decay may be an effect of the drop into energy gaps. It is difficult to judge whether the fast decay is intrinsic or not. At some other high energies, behavior of the conductance looks like the low energy cases. The result at $E = 1.52$ is shown in Fig.5. The fluctuations are large, but we might regard them as the power-law decays.

Finally, we remark on a special mechanism of scattering in QC, the scattering due to topological "disorder". As we mentioned, the number of sites changes from column to column in QC. When an incident wave is propagating in the scatterer consisting of a QC, some modes are perfectly reflected at the places where the site number is small because degrees of freedom are reduced there. Which modes are reflected depends on the local connectivity of sites. It is noted that this perfect reflection is irrespective of the energy of the incident wave. In this way, high resistivity in QC owes its origin to the topological "disorder".

References

[1] H. Tsunetsugu, T. Fujiwara, K. Ueda, and T. Tokihiro, J. Phys. Soc. Jpn. **55**, 1420 (1986).

[2] H. Tsunetsugu and K. Ueda, Phys. Rev. B **38**, 10109 (1988).

[3] Preliminary results of this work are published in, H. Tsunetsugu, J. Non-Cryst. Solids **117/118**, 781 (1990).

[4] D. Shechtman, I. Blech, D. Gratias, and J. W. Cahn, Phys. Rev. Lett. **53**, 1951 (1984).

[5] L. Bendersky, Phys. Rev. Lett. **55**, 1461 (1985).

[6] D. Levine and P. J. Steinhardt, Phys. Rev. Lett. **53**, 2477 (1984).

[7] For a review, P. A. Lee and T. V. Ramakrishnan, Rev. Mod. Phys. **57**, 287 (1985).

[8] E. Abrahams, P. W. Anderson, D. C. Licciardello, and T. V. Ramakrishnan, Phys. Rev. Lett. **42**, 673 (1979).

[9] M. Kohmoto, L. P. Kadanoff, and C. Tang, Phys. Rev. Lett. **50**, 1870 (1983).

[10] S. Ostlund, R. Pandit, D. Rand, H. J. Schellnhuber, and E. D. Siggia, Phys. Rev. Lett. **50**, 1873 (1983).

[11] M. Kohmoto, B. Sutherland, and C. Tang, Phys. Rev. B **35**, 1020 (1987).

[12] M. Kohmoto and B. Sutherland, Phys. Rev. Lett. **56**, 2740 (1986).

[13] K. Semba, Master Thesis, Tokyo Univ. (1985); K. Semba and T. Ninomiya, private communication.

[14] T. Fujiwara, M. Arai, T. Tokihiro, and M. Kohmoto, Phys. Rev. B **37**, 2797 (1988).

[15] M. Arai, T. Tokihiro, T. Fujiwara, and M. Kohmoto, Phys. Rev. B **38**, 1621 (1988).

[16] M. Gardner, Sci. Am. **236(1)**, 110 (1977).

[17] N. G. de Bruijn, Ned. Akad. Weten. Proc. Ser. A **43**, 39 (1981); **43**, 53 (1981).

[18] Only vertices K and S4 (names after Ref.[17]) correspond to even-member rings.

[19] V. Elser and C. L. Henley, Phys. Rev. Lett. **55**, 2883 (1985).

[20] O. Entin-Wohlman, M. Kléman, and A. Pavlovitch, J. Phys. (Paris) **49**, 587 (1988).

[21] P. W. Anderson, D. J. Thouless, E. Abrahams, and D. S. Fisher, Phys. Rev. B **22**, 3519 (1980).

[22] P. A. Lee and D. S. Fisher, Phys. Rev. Lett. **47**, 882 (1981).

[23] H. Tsunetsugu, unpublished.

Electronic Structure in Three-Dimensional Quasicrystals

T. Fujiwara and T. Yokokawa

Department of Applied Physics, University of Tokyo,
Bunkyo-ku, Tokyo 113, Japan

Abstract: Electronic structures in three-dimensional quasicrystals are discussed on the basis of those in approximated crystalline AlMn and AlCuLi alloys. The density of states in realistic alloys consists of a set of spiky peaks. The stability is mainly caused by the pseudogap at the Fermi energy in both alloys. The pseudogap is opened by a strong structural scattering. Centers of Mackay icosahedra in AlMn and those of triacontahedra in AlCuLi should be vacant from the viewpoint of stability.

1. Introduction

Electronic structures in quasicrystals are quite exotic. For example, the Fibonacci lattice, a one-dimensional correspondence of quasicrystals, has a purely singular continuous energy spectrum with zero Lebesgue measure and its wavefunctions are critical, *i.e.* neither extended nor localized.[1] The characteristics of critical wavefunctions can be clarified by the method of multifractals.[2] It was shown in two-dimensional Penrose lattices that the energy spectrum is ill-smooth and that wavefunctions are mostly power-law decaying.[3] The latter fact was confirmed by a finite size scaling study of the electric conductance in very large systems.[4] Infinitely degenerate states appear in simple tight-binding models, which are strictly localized within finite regions. We named them confined states and string states and gave analytical expressions of their degeneracy and wavefunctions.[5] In three-dimensional Penrose lattices, the density of states did not show any premonitory behavior of the singular spectrum, but still we cannot exclude the possibility that this might be just because of a limitation of the system size. Confined states were found also in three-dimensional systems.[6]

In the present paper, the electronic structures in crystalline analogues of realistic quasicrystals, AlMn and AlCuLi, will be presented and cohesive properties in quasicrystals will be discussed.

2. Periodic Penrose lattice

The three-dimensional Penrose lattice can be generated as the generalized grid method. The icosahedral basis vectors are chosen as $\mathbf{g}_\ell = (0,1,\tau)_{c.p.}$ for ℓ = 1,2,3 and $\mathbf{g}_\ell = (0,-1,\tau)_{c.p.}$ for ℓ = 4,5,6, where *c.p.* stands for cyclic permutation and τ is the golden mean $(\sqrt{5}+1)/2$. Then the hexagrids are defined as

$$G_6 = \{\mathbf{x} \in \mathrm{E}^3 \quad | \quad \mathbf{x}\cdot\hat{\mathbf{g}}_\ell - \gamma_\ell = k_\ell|\mathbf{g}_\ell|; \ell = 1,2,\cdot\cdot,6; k_\ell \in \mathrm{Z}\}, \tag{1}$$

where $\hat{\mathbf{g}}_\ell = \mathbf{g}_\ell/|\mathbf{g}_\ell|$. Regular intersection points in a grid space define prolate rhombohedra in the physical space. The icosahedral packings of the Penrose local isomorphism (PLI) class cannot be defined as a dual lattice of the hexagrids but by an introduction of four unit-cell shapes for singular intersection points. Four-, five- and six-fold singular

Springer Series in Solid-State Sciences, Vol. 93 **Quasicrystals**
Editors: T. Fujiwara · T. Ogawa © Springer-Verlag Berlin, Heidelberg 1990

intersection points correspond to a dodecahedron, a rhombic icosahedron and a rhombic triacontahedron, respectively.

Substituting an approximate rational number for the golden mean and using an appropriate γ_ℓ, we can get a split of the degeneracy of singular intersection points and a *periodic* Penrose lattice at a cost of phason introduction. The minimal rational approximation to the golden mean τ is $\tau_n = \mathrm{F}_{n+1}/\mathrm{F}_n$, where F_n is the n-th Fibonacci number defined by $\mathrm{F}_0 = 0$, $\mathrm{F}_1 = 1$ and $\mathrm{F}_n = \mathrm{F}_{n-1} + \mathrm{F}_{n-2}$. The periodic lattice is of body centered cubic symmetry for $n = 1, 4, 7, 10 \cdots$ and otherwise of simple cubic symmetry. We call them the (F_{n+1} / F_n)-periodic Penrose lattices. The period of a simple cubic cell is $(2+2/\sqrt{5})^{1/2}\tau^n$ in units of the length of an edge of a rhombohedron. The cubic cell consists of $4\mathrm{F}_{3n+2}$ prolate and $4\mathrm{F}_{3n+1}$ oblate rhombohedra.[6]

3. Simple tight-binding models [6]

Before going to realistic alloys, we will give a summary of the results of simple tight-binding models in three-dimensional Penrose lattices. Two tight-binding models can be defined in a three-dimensional Penrose lattice, when we assume one s-orbital per atom. Atoms are placed either on centers of rhombohedra with constant hopping interactions (-1) between adjacent rhombohedra (a center model) or on vertices with constant hopping interactions (-1) along edges of rhombohedra (a vertex model). In both cases, the discrete Schrödinger equation can be given as

$$\sum_j t_{ij}\psi_j = E\psi_i, \tag{2}$$

where $t_{ij} = -1$ for an interacting pair (i, j) and 0 otherwise. The density of states (DOS) did not show any premonitory behavior of a singular spectrum as in a two-dimensional Penrose lattice possibly due to a limitation of the system size. The shape of the DOS is, in the vertex model, similar to that of the simple cubic lattice, because the lower order moments of the spectrum are the same. The degeneracy was found at the energy $E = 2$ in the center model and at $E = 0$ in the vertex model. The wavefunctions of these states are strictly localized in the sense that their amplitude becomes abruptly zero outside some finite region, and have a constant amplitude on some specific sites with alternating signs. We thus call these states confined states. We can find the same tiling configurations supporting the confined states distributing uniformly in a system. Therefore, the degeneracy of the confined states is proportional to the system size and is infinite in an infinite three-dimensional Penrose lattice.

4. AlMn alloy

4-1. Structure [7, 8]

Crystalline α(AlMnSi) is the three-dimensional (1/1)-periodic Penrose lattice. The golden mean τ is approximated to τ_1. The unit cell is a packing of six rhombic dodecahedra (RDs) and eight prolate rhombohedra (PRs), shown in Fig.1. Atomic decoration is fulfilled only in a RD and atoms on its faces are shared with PRs. In the α(AlMnSi) structure, there are six classes of atomic sites; one Mn atomic site and five Al or Si atomic sites (α, $\beta1$, $\beta2$, γ and δ). Large holes are at (0,0,0) and $(\frac{1}{2}, \frac{1}{2}, \frac{1}{2})$, which are the vertex positions of a RD connected by the longest body diagonal. Around this hole (MI center), 12 Al(α), 12 Mn, 24 Al($\beta1$) and 6 Al($\beta2$) form the 54-atom Mackay icosahedron (MI). The Mn are on the vertices of a MI. The Al(α) are each at the midpoint between a hole (MI center) and a Mn and form an inner icosahedron. Two adjacent MIs (in the

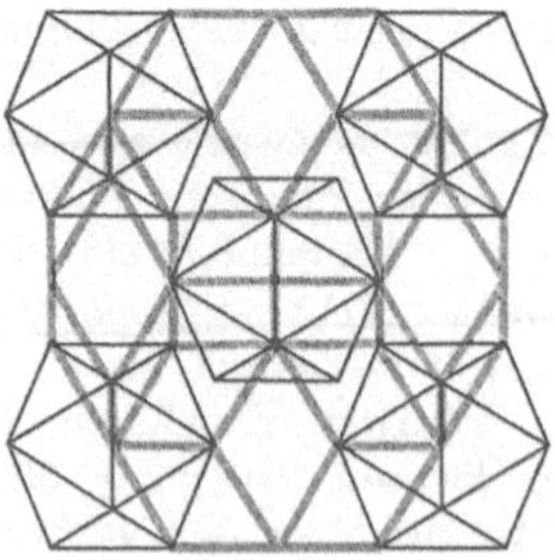

Fig.1. Idealized α(AlMnSi) structure consisting of Mackay icosahedra projected onto the (100)-plane. The shaded lines show five rhombic dodecahedra.

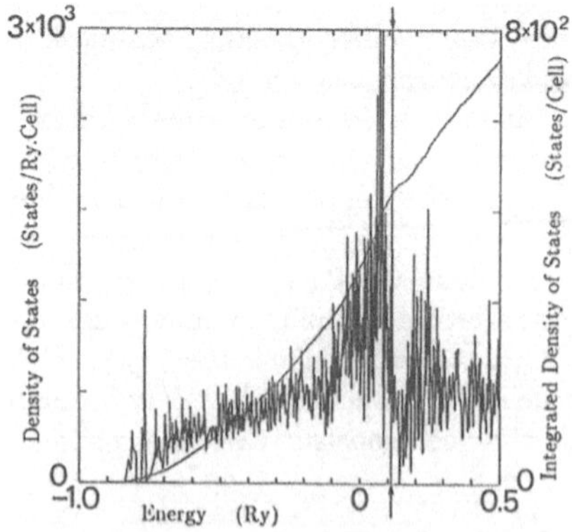

Fig.2. Density of states and integrated density of states in α(AlMnSi). Fermi energy is indicated by the vertical line and arrows.

(111) direction) share a distorted Mn octahedron. A Mn-Mn nearest neighbor contact is not preferable. The Al(β1) are on the edges of a MI (at the midpoint between two Mn atoms) and are shared with a Mn octahedron. The Al(β2) are each on one of other six edges of a MI (at the midpoint between two Mn atoms). All γ-sites (24 sites) and half of the δ-sites (6 sites in one bcc-sublattice) are occupied by Al or Si atoms, which are glue atoms for a MI pair in the (100) direction. The other half of the δ-sites (6 sites in the other bcc-sublattice) are unoccupied. Therefore, α(AlMnSi) is actually a simple cubic crystal and the space group is *Pm*3.

4-2. Electronic structure of the AlMn alloy [9, 10]

The electronic structures of α(AlMn) were calculated self-consistently by the *linear muffin-tin orbital–atomic sphere approximation* method [11] with the local density functional theory. In vacant spatial regions, such as centers of MIs and PRs, fictitious vacant atoms with zero atomic number were placed. A total of 154 atoms, 114 Al, 24 Mn and 16 vacant atoms, were used in a simple cubic unit cell ($Al_{82.6}Mn_{17.4}$). Al atoms were substituted for Si atoms in the calculation because favorable positions of Si atoms in AlMnSi have not been experimentally distinguished from those of Al atoms.

The total DOS and the integrated DOS are shown in Fig.2, where the Fermi energy E_F is 0.115 Ry. The DOS consists of a set of spiky peaks; every band is highly dispersionless. This is not due to the localization of wavefunctions and the character of a wavefunction in one band should depend strongly on the momentum **k**. The main peak at $E \simeq 0.0 \sim 0.1$Ry, which originates from the Mn 3d-states, is split into two at E_F and the other peak locates at 0.22Ry. This splitting and opening of a pseudogap at $E \simeq 0.14$Ry causes a deviation of the high DOS peak from the Fermi energy E_F, which relatively stabilizes the system. With a few Si atoms substituted for Al atoms, the Fermi energy climbs towards the minimum of the pseudogap and the stability would be much enhanced. The deep tail of the DOS is mainly of Al 3s character. If we assume the free electron energy dispersion, the minimum of the pseudogap measured from the bottom of the Al s·p band corresponds to a momentum $k = 1.75\text{Å}^{-1}$. The value $2k$ fairly coincides with a momentum transfer Q of strong diffraction spots in the icosahedral Al-Mn alloy, the (211111) spot with $Q = 2.896\text{Å}^{-1}$ (corresponding to $(0\bar{3}5)$ in the α-phase) or (221001) with $Q = 3.043\text{Å}^{-1}$ ((006) in the α-phase).[8]

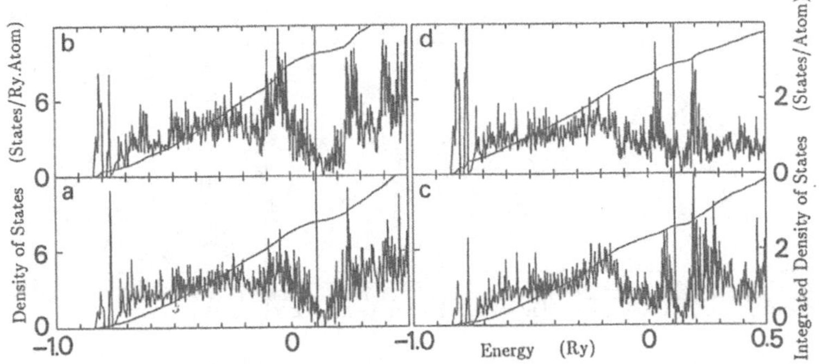

Fig.3. Local DOS's of (a) Al(β1), (b) Al(β2), (c) Al(γ) and (d) Al(δ) in the bcc-sublattice with δ-sites occupied by Al atoms.

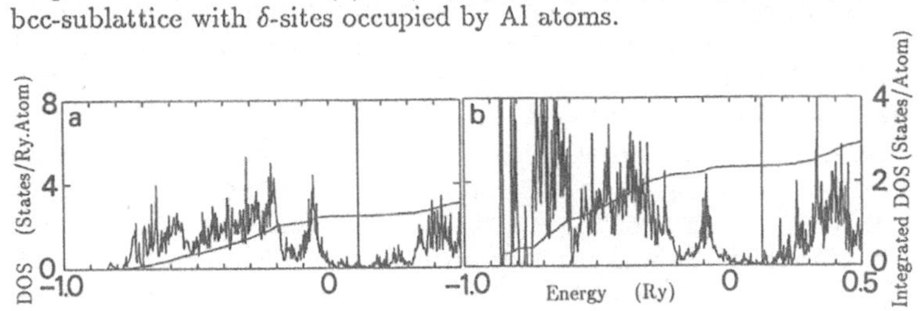

Fig.4. Local DOS of MI centers; (a) vacant center, (b) center occupied by an Al atom.

All Al sites have, to some extent, similar projected DOS. The local DOS of Al(β1), Al(β2), Al(γ) and Al(δ) in the bcc-sublattice with δ sites occupied by Al atoms are shown in Fig.3. The DOS peak feature of Al(α), Al(β1) and Al(β2) is quite common in transition metal compounds, in which the s·p-states are hybridized with the low energy tail of transition metal d-states but reduced (antiresonance dip) in an energy range of the main Mn d-peak. On the other hand, in Al(γ) and Al(δ), the DOS in a tail range of Mn d-states($-0.1 \sim 0.05$Ry) is reduced but has a peak resonated with the Mn d-state at E$\simeq 0.05 \sim 0.10$Ry, and this fact suggests that bonding Mn d-electron lobes point to glue Al atoms and form strong bonds. Other strongly bonding states can be seen at the bottom of the band (E$\simeq -0.8$Ry) in Al(β2) and glue Al atoms. These two facts are typical features of *glue* atoms.

A MI center shows no DOS in a wide energy range near the Fermi energy ($-0.05 \sim 0.2$Ry). The center of a MI has an enough space to capture an Al atom but actually it remains vacant. We calculated a self-consistent electronic structure again in a fictitious system with Al atoms occupying the MI centers.[9, 10] General features have not altered very much except the enhancement at E$\simeq -0.6 \sim -0.8$Ry in the DOS of Al (MI center), as compared in Fig.4. The local DOS of Al (MI center) vanishes in an energy range around the Fermi energy exactly as on the original vacant MI center. The calculated total valence charge on Al (MI center) is about 2.5e/atom, the remaining excess valence charge is distributed on the surrounding atoms at the Fermi energy and no energetic stability is gained. Therefore, the MI center should remain vacant.

5. AlCuLi alloy

5-1. Structure [12, 13]

Al_5CuLi_3 (R-phase) is the crystalline analogue of an icosahedral AlCuLi alloy (T2-phase) and is the three-dimensional (1/1)-periodic Penrose lattice. There are eight

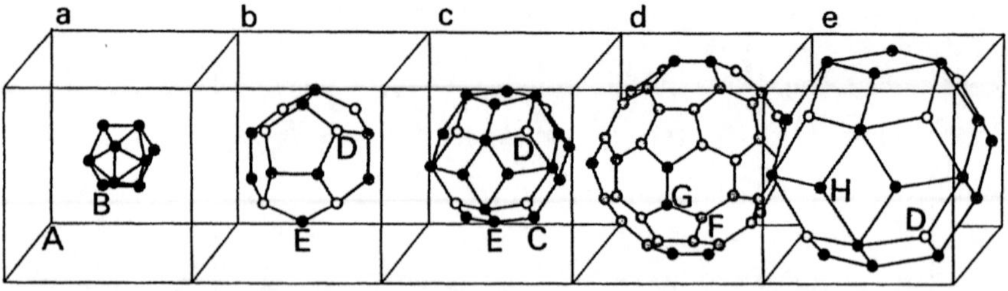

Fig.5. Shell structure of R(AlCuLi) ((a)-(e)). For details see the text.

different classes of atomic sites, A-H. Sites A are vertices of a PR or a RD, connected by the longest body diagonal. Sites C are other vertices of a PR or a RD. A midpoint of each edge is a B or a F site. Two sites D are inside a PR, and eight sites 4 E, 2 H and 2 G are in a RD. Totally the numbers of sites in a simple cubic unit cell are 2 A, 24 B, 24 C, 16 D, 24 E, 48 F, 12 G and 12 H.

Figure 5 shows the R(AlCuLi) bcc structure: (a) first an icosahedron of B centered at a vacant A site (0,0,0) or $(\frac{1}{2},\frac{1}{2},\frac{1}{2})$, (b) a pentagonal dodecahedron of 8 D and 12 E above the faces of the icosahedron of B, (c) a smaller triacontahedron (s-TCH) of 12 C (capped on face-centers of a pentagonal dodecahedron (b)), 8 D (appearing already in (b)) and 12 E (appearing already in (b)), (d) a distorted truncated triacontahedron (t-TCH) of 48 F (a hexagonal face of F is shared with a nearest neighbor t-TCH,) and 12 G (located on the faces of a simple cubic unit cell and shared with a next nearest t-TCH), (e) a larger triacontahedron (ℓ-TCH) of 24 H and 8 D (appearing already in (b)). The ℓ-TCHs share a rhombic face of H with each other in the (100) direction and an overlapping oblate rhombohedron of 6 H and 2 D in the (111) direction. The structure does not contain a glue atom once it is regarded as a packing of touching and overlapping ℓ-TCHs. Sites A, centers of icosahedra of B, are vacant. Sites D, E and H are occupied by Li atoms, and sites G by Al. Sites B, C and F are occupied randomly by Al and Cu and their populations (Al/Cu) are 88.6/11.4, 53.8/46.2 and 88.6/11.4, respectively. The space group is $Im3$ of bcc. [13]

5-2. Electronic structure of the AlCuLi alloy

Vacant atoms are placed on A sites. The calculation was done in a model structure, where Al atoms were placed on sites B and F, Cu atoms on sites C of s-TCHs in one bcc-sublattice and Al atoms on sites C of s-TCHs in the other bcc-sublattice. In total, 162 atoms, 96 Al, 12 Cu, 52 Li and 2 vacant atoms, were placed in a simple cubic unit cell ($Al_{60.0}Cu_{7.5}Li_{32.5}$). The symmetry of the model system is not bcc, but simple cubic.

The total DOS and the integrated DOS are shown in Fig.6, where the Fermi energy E_F is 0.051Ry. Just as in the AlMn alloy, the DOS consists of a set of spiky peaks. The main peak at E$\simeq -0.3 \sim -0.1$Ry originates from Cu 3d-states and is not split. The DOS shows a clear pseudogap at the Fermi energy. We substitute Al atoms for Cu atoms, the position of the pseudogap measured from the bottom of the valence band does not change at all, but the Fermi energy shifts to the higher energy side because of the increasing total number of effective valence electrons.[14] Therefore, the Cu d-states do not contribute to the formation of the pseudogap or strong bonding orbits. The main role of Cu atoms is the supply of valence electrons and the Cu concentration of a stable system should be determined so that the Fermi energy may be located just at the bottom of the pseudogap. The pseudogap and appropriate contents of Cu atoms energetically stabilize the system. The deep tail of the DOS originates from Al and Li s-states. The Fermi energy measured from the bottom of the band corresponds to

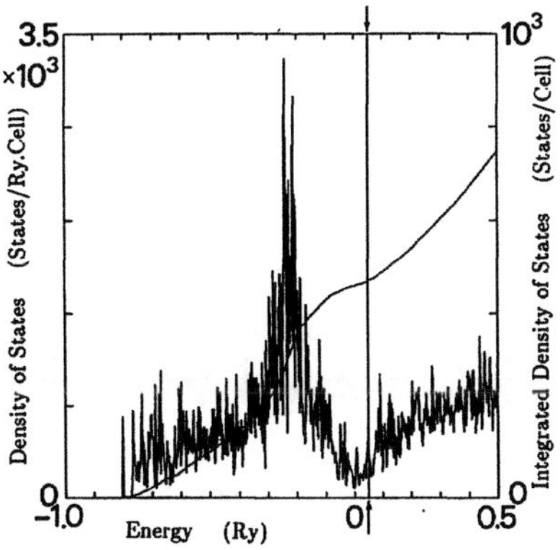

Fig.6. Density of states and integrated density of states in R(AlCuLi). Fermi energy is indicated by the vertical line and arrows.

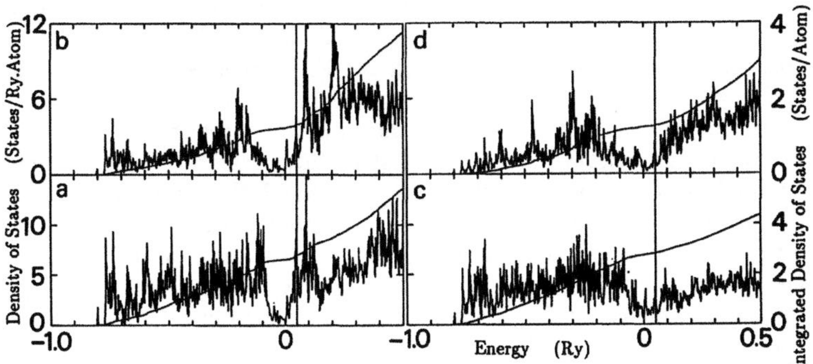

Fig.7. Local DOS of (a)Al on G site, (b)Li on H site, (c)Al on F site, and (d)Li on D site, in the bcc-sublattice with C sites occupied by Cu atoms.

a momentum $k = 1.64\text{Å}^{-1}$. The strong diffraction spots in an icosahedral AlCuLi are the (221001) spot with a momentum transfer $Q = 2.793\text{Å}^{-1}$ (corresponding to (253) or (006) in the R-phase) and the (222100) with $Q = 3.189\text{Å}^{-1}$ ((710) in the R-phase).[15] Coincidence of $2k$ and Q is again observed.

An A site (vacancy) has almost zero DOS in the energy range around the Fermi energy ($-0.2 \sim 0.2$Ry). We tried to calculate the electronic structures in the case that centers of icosahedra (A sites) are occupied by Al atoms, but could not obtain self-consistently converged results. This may indicate that the A sites should remain vacant from the viewpoint of energetic stability. Electron charge flows from surrounding B sites into a vacant A site, forms several bonding states in an energy range $-0.8 \sim -0.5$Ry and shows narrow bonding peaks of the DOS. The local DOS of the A and B sites in the bcc-sublattice with Cu atoms on C sites is shifted by 0.1Ry to the higher energy side comparing with those in the bcc-sublattice with Al atoms on C sites. The values of an electrostatic potential on A and B sites are sensitive to the surrounding charge density on C sites.

Figure 7 shows the local DOS in R(AlCuLi). Different specific characters can be found in two groups: (G, H) sites and (D, F) sites. The pseudogap at E_F is much clearer in the former group, (G, H) sites, than the latter group (D, F). The DOS of G and H sites have a narrow peak at $E \simeq 0.1$Ry and the Fermi energy in the present case is just below this peak. Therefore, once the number of valence electrons increases, the Fermi energy climbs up this peak and the system may suffer from energetic instability. As to the ℓ-TCH, the G and H sites are shared by a pair in the (100) direction. The sites D and F (and H) are shared by a pair of ℓ-TCH in the (111) direction. The total number of valence electrons is, therefore, strongly correlated with the cluster geometry in the (100) or (111) directions. The structural stability seems to be enhanced, in the present situation, in the (100) direction more than in the (111) direction.

6. Discussion

We have briefly reviewed the tight-binding models in the three-dimensional Penrose lattice and presented electronic structures in crystalline analogues of icosahedral alloys AlMn and AlCuLi.

The clusters, Mackay icosahedra in AlMn or triacontahedra in AlCuLi, are constituent units of three-dimensional Penrose lattices and placed on twelve-fold vertices of the quasiperiodic lattice. They have chemical bonds or a strong hybridization within a cluster unit and are quite stable. The central sites of both cluster units were shown to be vacant from the viewpoints of energetic stability. In both realistic alloy systems AlMn and AlCuLi, wide pseudogaps are opened at the Fermi energies and cause relative stability of the systems. Their positions measured from the bottom of the valence band coincide with the momentum transfers of the strong diffraction spots and the origin of the pseudogaps is ascribed to the strong structural scattering. These strong diffractions are caused by atomic planes with higher indices and, therefore, intermediate range correlation of cluster units (Mackay icosahedra in AlMn or rhombic triacontahedra in AlCuLi) is quite important. We might say that the stabilization mechanism is structural in the intermediate range and should be just the same in both quasicrystals and crystals. The stoichiometry is an essential factor determining the structure of the system, quasicrystalline or crystalline. The calculated DOS at the Fermi energy in AlMn is about 0.43 states/eV·atom and that in AlCuLi is 0.12 states/eV·atom. The difference of the DOS values at the Fermi energy in crystalline phases and the widths of the pseudogaps may relate to the stability difference of these two icosahedral alloys: metastable icosahedral AlMn and stable icosahedral AlCuLi. In AlCuLi, we obtain more information about the rule for the cluster aggregation, *i.e.* the ℓ-TCH may be joined favorably face-to-face, because the ℓ-TCH pair in the (100) direction is joined in this way and shows a more pronounced pseudogap than the pair in the (111) direction. However, in AlCuLi, the geometry of the adhering triacontahedra is quite sensitive to the total number of valence electrons. Once the number of valence electrons increases, the stability of the (111) directional pair should be enhanced.

The present calculation might explain the variety of the stability of AlCuTM phases such as glassy (TM=Ti and V), metastable icosahedral (TM=Cr and Mn), stable icosahedral (TM=Fe), stable decagonal (TM=Co) and crystalline (TM=Ni).[16] For the transition metal elements on the left hand side of the periodic table, the Fermi energy might sit on the peak position of the transition metal d-band. For the elements of the right hand side with increasing total number of valence electrons, the Fermi energy might shift to the minimum of the DOS.

The small cluster calculation of icosahedral Al_{13} and Al_{33} shows that the Fermi energy sits on the highest peak of the DOS.[17] The highest occupied molecular orbitals in these clusters are spatially very delocalized, of p$-\pi$ character. The substitution of a central Mn atom for an Al atom causes the p$-\pi$ bonding orbitals to empty and the localized p$-$d antibonding orbital to be occupied. However, this is not really the case in condensed phases and these small cluster calculations cannot explain the structural stability. As stated above, the results of the present calculation show an important unique role of the intermediate range cluster-cluster correlation in stabilizing quasicrystalline and crystalline materials.

The DOS value at the Fermi energy can be determined experimentally by the coefficient of the linearly temperature dependent term of the specific heat. Several authors have mentioned the role of s-d resonance and the DOS enhancement at the Fermi energy in icosahedral AlMn.[18] However, as clearly shown in Fig.2, the Fermi energy is in the pseudogap. In AlMn alloys, the measured values are dominated by an uncertainty due to its spin-glass-like behavior.[18] On the other hand, the reduced DOS values at the Fermi energy in AlCuLi and other stable non-magnetic quasicrystals seem to be a common conclusion.[19, 20, 21] The measured DOS at the Fermi energy of non-magnetic metastable quasicrystals is larger than that in stable quasicrystals.[20] Our result gives theoretical support to these conclusions, *i.e.* the reduced DOS in stable quasicrystals, and presents a proposal that this is also common in metastable quasicrystals.

Acknowledgements

The present work is partly supported by a Grant in Aid from the Japan Ministry of Education, Science and Culture. We have benefited from fruitful discussions with S.Takeuchi. We also wish to thank S.Takeuchi, U.Mizutani and S.Matsuo for sending preprints before publication.

References

1 P.A.Kalugin, A.Yu.Kitaev, and L.S.Levitov, Zh.Eksp.Theor.Fiz. **91**, 692 (1986) [Sov.Phys. - JETP **64**, 410 (1986)];
M.Kohmoto, B.Sutherland, and C.Tang, Phys.Rev. B**35**, 1020 (1987).

2 T.Fujiwara, M.Kohmoto and T.Tokihiro, Phys.Rev. B**40**, 7413 (1989);
M.Kohmoto, in the present volume.

3 H.Tsunetsugu, T.Fujiwara, K.Ueda, and T.Tokihiro, J.Phys.Soc.Jpn. **55**, 1420 (1986).

4 H.Tsunetsugu, in the present volume.

5 T.Fujiwara, M.Arai, T.Tokihiro, and M.Kohmoto, Phys.Rev. B**37**, 2797 (1988); M.Arai, T.Tokihiro, T.Fujiwara, and M.Kohmoto, Phys.Rev. B**38**, 1621 (1988); T.Tokihiro, T.Fujiwara, and M.Arai, Phys.Rev. B**38**, 5981 (1988); T.Fujiwara, T.Tokihiro and M.Arai, Materials Science Forum **22-24**, 459 (1987).

6 M.Krajci and T.Fujiwara, Phys.Rev.B**38**,12903 (1989).

7 V.Elser and C.L.Henley, Phys.Rev.Lett. **55**, 2883 (1985).

8 M.Audier and P.Guyot, Phil.Mag. B**53**, L43 (1986).

9 T.Fujiwara, Phys.Rev.B**40**, 942 (1989).

10 T.Fujiwara, J. Non-Crystal. Solids (*Proc. 7th Int. Conf. Liquid and Amorphous Metals*, September 4-8, 1989, Kyoto, Japan).

11 O.K.Andersen, Phys.Rev.B**12** (1975) 3060.

12 C.L.Henley and V.Elser, Phil.Mag. B**53**, L59 (1986).

13 M.Audier, J.Pannetier, M.Leblanc, C.Janot, J-M.Lang and B.Dubost, Physica B**153**, 136 (1988).

14 T.Yokokawa and T.Fujiwara, unpublished work.

15 M.A.Marcus and V.Elser, Phil.Mag. B**54**, L101 (1986); M.Audier, P.Sainfort and B.Dubost, Phil.Mag. B**54**, L105 (1986).

16 A.Inoue, in the present volume.

17 M.E.McHenry, M.E.Eberhart, R.C.O'Handley and K.H.Johnson, Phys.Rev.Lett. **56**, 81 (1986).

18 C.Berger, J.C.Lasjaunias, J.L.Tholence, D.Pavuna and P.Germi, Phys.Rev.B**37**, 6525 (1988).
19 K.Kimura, H.Iwahashi, T.Hashimoto, S.Takeuchi, U.Mizutani, S.Ohashi and G.Itoh, to be published in J.Phys.Soc.Jpn.
20 J.L.Wagner, K.M.Wong and J.Poon, Phys.Rev.B**39**, 8091 (1989).
21 S.Matsuo, H.Nakano, T.Ishimasa and Y.Fukano, to be published in J.Phys.: Condensed Matters; U.Mizutani, Y.Sakabe and T.Matsuda, to be published in J.Phys.: Condensed Matter.

Part VI

Spin Statistics

Monte Carlo Study of Spin Statistics in Two- and Three-Dimensional Quasicrystals

Y. Okabe and K. Niizeki

Department of Physics, Tohoku University, Sendai 980, Japan

Abstract. The ferromagnetic Ising model on the three-dimensional icosahedral quasilattice is studied by using the Monte Carlo simulation. We treat both cases where the spins are located on the vertices of the lattice and the centers of the lattice. Investigating the critical phenomena on the basis of finite-size scaling, we confirm the universality of critical exponents. The frustration effect is studied in the antiferromagnetic Ising model on the dual two-dimensional Penrose lattice. We find the growth of the fluctuation of sublattice magnetization at low temperatures.

1. Introduction

Since the discovery of quasicrystals, growing attention has been given to the spin statistics in quasicrystals. The spin systems on the one-dimensional (1D) quasilattice can be solved with an exact renormalization approach [1], and interesting behavior which comes from the characteristic of quasilattice was pointed out. It is more interesting to investigate the role of the quasiperiodicity and the self-similarity in phase transition and critical phenomena. To do so, however, higher-dimensional systems should be studied.

The Ising model on the 2D Penrose lattice was studied by means of an approximate renormalization scheme [2]. Some special model was considered with an exact treatment [3]. The Monte Carlo simulation is a powerful method to obtain precise numerical information. The present authors [4] made a Monte Carlo study of the Ising model on the Penrose lattice of a size up to 439,204 sites. Other authors [5-8] also studied the Ising model by use of Monte Carlo technique for smaller systems. Quite recently, the high-temperature expansion [9] and the coherent anomaly method [10] have been applied to the study of the Ising model on the

Springer Series in Solid-State Sciences, Vol. 93 **Quasicrystals**
Editors: T. Fujiwara · T. Ogawa

Penrose lattice. We should note that the Potts model on the Penrose lattice was also studied by Monte Carlo simulation [11].

In the previous Monte Carlo study of the 2D Penrose lattice [4], we showed the universality of critical exponents and the duality between the critical temperatures of dual lattices. We also pointed out that the critical temperature for the Penrose (dual Penrose) lattice is very close to that for the dice (Kagomé) lattice. A similar behavior was also found in the study of the percolation threshold for the bond percolation problem of the Penrose and its dual lattices [12]. We can show that the high-temperature series coefficients of the Penrose lattice are numerically close to those of the dice lattice. This similarity of the local structure may give the reason for the closeness of the critical temperatures of the Penrose lattice and of the dice lattice.

It is interesting to study spin systems on more realistic 3D quasilattices. In the first part of this paper, we report the extension of the Monte Carlo study to the 3D icosahedral lattice. Second, we investigate the antiferromagnetic Ising model on the 2D dual Penrose lattice, paying attention to the frustration effect in quasicrystals. This system does not have long-range order due to a frustration.

2. Ferromagnetic Ising Model on the Icosahedral Quasilattice

The first topic is the ferromagnetic Ising model on the 3D icosahedral lattice. We treat both cases that the spins are located on the vertices of the rhombohedron and the centers of the rhombohedron. In the former case, the coordination number z ranges from 4 to 12. The average coordination number is 6, which is the same as the simple cubic lattice. From now on, we simply call this system the icosahedral lattice. On the other hand, the coordination number is fixed to be 6 for the latter case. The latter system may be regarded as a 'dual' lattice of the icosahedral lattice. But we should note that it is *not* dual in the sense that the Euler theorem holds. Numerical simulations are performed on finite systems; therefore, it is desired to make the size effect small. For this purpose we make simulations on the 'periodic' icosahedral lattice with a large unit cell [13]; the golden ratio $\tau = (1+\sqrt{5})/2$ is approximated by a consecutive pair of Fibonacci numbers.

We are concerned with the Ising Hamiltonian

$$\mathcal{H} = -J \sum_{<i,j>} \sigma_i \sigma_j, \qquad (\sigma_i = \pm 1). \tag{1}$$

We follow the usual Metropolis method of the Monte Carlo simulation. For the FORTRAN coding, we extend the fast algorithm of multispin coding by Bhanot et al.[14]. This algorithm has been successfully used to investigate the critical phenomena of the 3D Ising model [15]. For the icosahedral lattice, the coordination number ranges from 4 to 12; the energy difference at a spin-update trial takes 25 values. Although the logical operations become complicated, it is straightforward to apply the multispin coding of Bhanot et al.[14]. It is known that the calculation is fully vectorized if the lattice is decomposed into appropriate interpenetrating sublattices; such a sublattice decomposition is simply guaranteed for the icosahedral lattice because it is a bipartite lattice. With these techniques we can realize the speed of the Monte Carlo simulation of the same order as that for the regular lattice. The coding of simulation is much simpler for the 'dual' icosahedral lattice because all the coordination numbers are fixed to be 6, which is the same as the simple cubic lattice. It should be noted that the 'dual' icosahedral lattice can be decomposed into ten interpenetrating sublattices; the vectorization of the calculation is also possible in this case.

We calculated the square of the magnetization per spin $< m^2 >$. In Fig. 1, we show the temperature dependence of $\sqrt{< m^2 >}$ for the ferromagnetic Ising model on the icosahedral lattice and on its 'dual' lattice. The lattice sizes are shown in the inset. The largest lattice has 43,784 sites. Simulations were made for 10,000 Monte Carlo steps per spin (MCS), and the first 2,000 MCS were excluded when taking an average. Two independent runs were made for each size and temperature. In the limit $N \to \infty$, the quantity $\sqrt{< m^2 >}$ yields the spontaneous magnetization for $T < T_c$, and $< m^2 >$ becomes the susceptibility $\chi T/N$ for $T > T_c$. From Fig. 1, we can clearly see the second-order paramagnetic-ferromagnetic phase transition for both quasilattices. The critical temperature for the simple cubic lattice is shown by an arrow, which is lower than T_c's for both quasilattices. This point will be discussed later.

In order to study the critical properties in detail, we made longer-time simulations for the temperatures near T_c. The temperature depen-

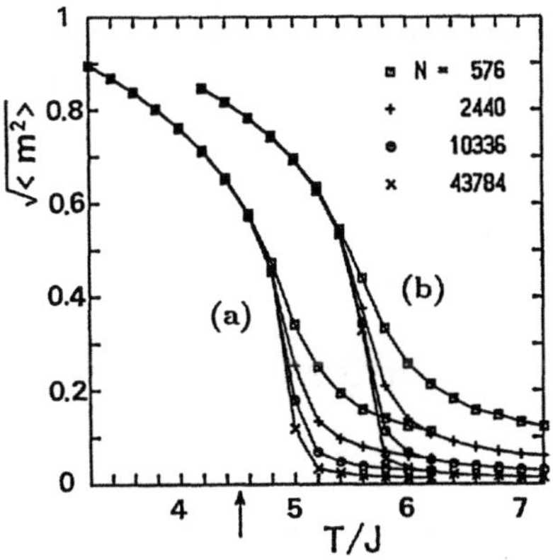

Fig. 1. Temperature dependence of $\sqrt{< m^2 >}$ for the ferromagnetic Ising model on the icosahedral lattice (a) and its 'dual' lattice (b). The lattice sizes are shown in the inset. The arrow gives T_c for the simple cubic lattice.

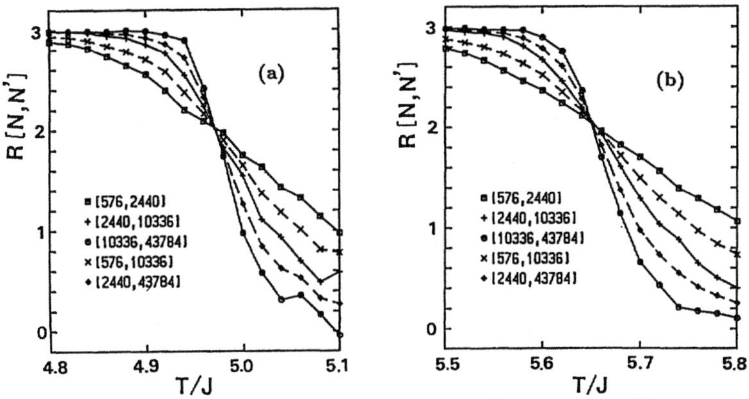

Fig. 2. Temperature dependence of the ratio of $< m^2 >$ for the icosahedral lattice (a) and the 'dual' icosahedral lattice. The quantity $R[N, N']$ is defined by eq. (2). The sizes of pairs, $[N, N']$, are given in the inset.

dence of $< m^2 >$ for $4.8 \leq T/J \leq 5.1$ for the icosahedral lattice and for $5.5 \leq T/J \leq 5.8$ for the 'dual' icosahedral lattice were investigated; simulations were made for 400,000 MCS, and the first 80,000 MCS were discarded when taking an average. To estimate the critical temperature and the critical exponent, we use a phenomenological Monte-Carlo-renormalization-group analysis due to Barber and Selke [16]. We examine the ratio of the values of $< m^2 >$ for different sizes. We plot the temperature dependence of the quantity

$$R[N, N'] = \frac{\log(< m^2 >_L / < m^2 >_{L'})}{\log(L/L')} + d \qquad (2)$$

for both quasilattices in Fig. 2, where the linear size L is given by $\sqrt[3]{N}$,

Table I. The estimated T_c and $2y_H - d$ for the icosahedral lattice and the 'dual' icosahedral lattice. For comparison we also give the data for the simple cubic lattice.

lattice	T_c	$2y_H - d$
icosahedral	4.972 ± 0.06	2.00 ± 0.05
'dual' icosahedral	5.650 ± 0.08	2.01 ± 0.05
simple cubic	4.512	1.97

and $d = 3$. Obviously, $R \to 0$, for $T \to \infty$, and $R \to d$ for $T \to 0$. If corrections to finite-size scaling are negligible, all pairs $[N, N']$ are expected to intersect at a single point; its abscissa gives the critical temperature T_c and its ordinate the magnetic exponent $2y_H - d (= \gamma/\nu)$. From Fig. 2, we can safely estimate the critical temperature and the magnetic exponent for both quasilattices. In Table I, the estimated T_c and $2y_H - d$ are tabulated.

The obtained critical exponent y_H has no difference from that of the 3D regular lattices within the errors of simulation. We again confirmed the universality of critical exponents for 3D quasilattices. The estimated T_c's for both quasilattices are higher than that of the simple cubic lattice, $T_c/J = 4.512$, although the average coordination numbers of the these three lattices are the same. In the 2D case, the critical temperature for the square lattice is in the middle of the Penrose lattice and the dual Penrose lattice because of the duality relation,

$$\sinh(2J/T_c) \sinh(2J/T_c^*) = 1. \tag{3}$$

In the present 3D case, this duality relation is *not* satisfied because the Euler theorem does not hold in our 'dual' lattice. There was a conjecture that the lattice which has heterogeneity in z has a higher T_c than the homogeneous lattice with the same z for 2D lattices [17]. It is *not* the case for 3D lattices because the T_c for the 'dual' icosahedral lattice is higher than that for the icoshedral lattice. The high-temperature series may help the study of the local structure of the lattice, i.e., the critical temperature.

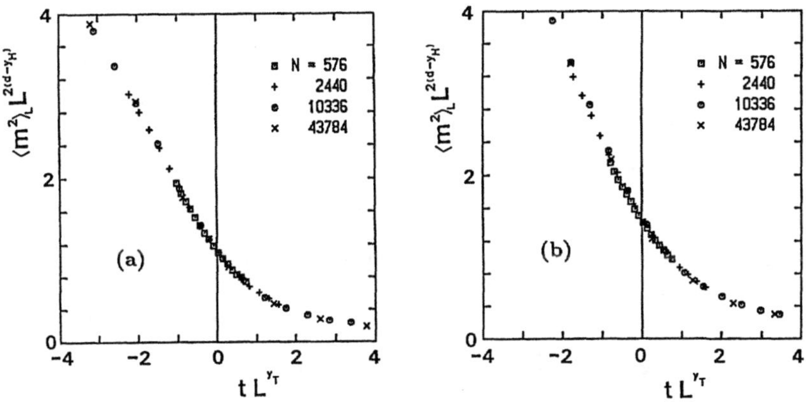

Fig. 3. Finite-size scaling plot; $< m^2 >_L L^{2(d-y_H)}$ versus tL^{y_T} with $t = (T-T_c)/T_c$. The three parameters are chosen as $T_c = 4.975$, y_H=2.487 and y_T=1.51 for the icosahedral lattice (a) and T_c = 5.655, y_H=2.487 and y_T=1.51 for the 'dual' icosahedral lattice (b).

In Fig. 3, we show a finite-size scaling plot

$$< m^2 >_L L^{2(d-y_H)} = f(tL^{y_T}), \quad t = (T - T_c)/T_c. \tag{4}$$

There are three fitting parameters, T_c, y_H and y_T, in this plot. The thermal exponent y_T is the inverse of the correlation length exponent; $y_T = 1/\nu$. Figure 3 shows that choosing $T_c = 4.975$, $y_H = 2.485$ and y_T = 1.59 gives a fairly good scaling plot for the icosahedral lattice, where we have used the critical exponents of 3D regular lattices. In the case of 'dual' icosahedral lattice, we chose T_c=5.655 and the same parameters for critical exponents.

3. Antiferromagnetic Model on the 2D Dual Penrose Lattice

In this section we study the antiferromagnetic Ising model on the 2D dual Penrose lattice, paying attention to the frustration effect. Figure 4 gives the Monte Carlo result of the temperature dependence of the susceptibility. At low temperatures the susceptibility diverges as $\sim 1/T$. There exists no long-range order in this model due to a frustration. The situation is similar to the case of the antiferromagnetic Ising model on the

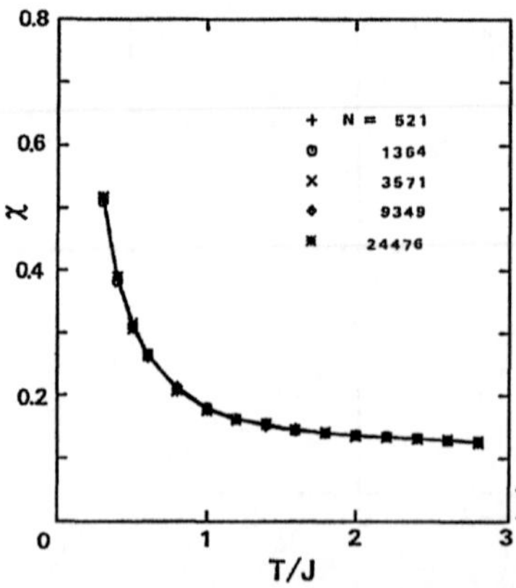

Fig. 4.

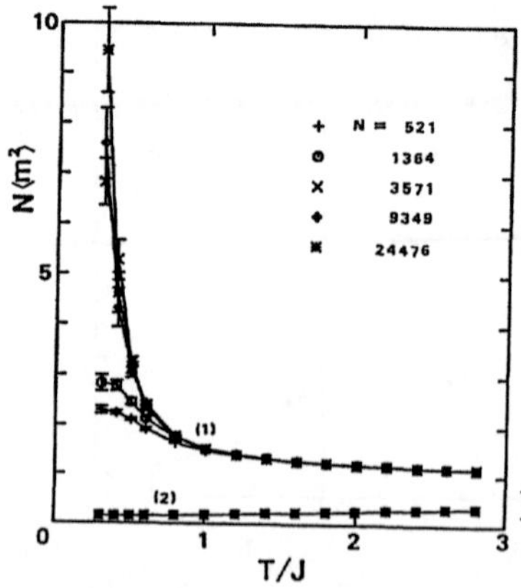

Fig. 5.

Fig. 4. Temperature dependence of the susceptibility for the antiferromagnetic Ising model on the dual 2D Penrose lattice.

Fig. 5. Temperature dependence of the fluctuation of the five-sublattice magnetization (1) and that of the total magnetization (2) for the antiferromagnetic Ising model on the dual 2D Penrose lattice.

triangular lattice. The triangular lattice can be decomposed into three sublattices, and the fluctuation of the sublattice magnetization grows at low temperatures. Inspired by this fact, we examined the sublattice magnetization for the dual Penrose lattice. Here we consider the decomposition into ten sublattices which are associated with obtuse rhombi with five orientations and acute rhombi with five orientations. We may group a set of sublattices of obtuse and acute rhombi perpendicular to each other into one sublattice. We show the square of the fluctuation of this five-sublattice magnetization in Fig. 5. For comparison, we also show the fluctuation of the total magnetization. From this figure, we find the growth of the fluctuation of the sublattice magnetization at low temperatures in the dual Penrose lattice.

In this system it is expected that the system becomes unstable against a small perturbation such as a uniform magnetic field. The ground-state configuration of spins as a function of magnetic field can be easily obtained for a regular lattice. In the case of a quasilattice, however, the exact configuration is difficult to obtain even at the ground state. Here, we calculate the 'ground' state under the condition that all the spins on each ten sublattices take the same orientaion. Figure 6 gives

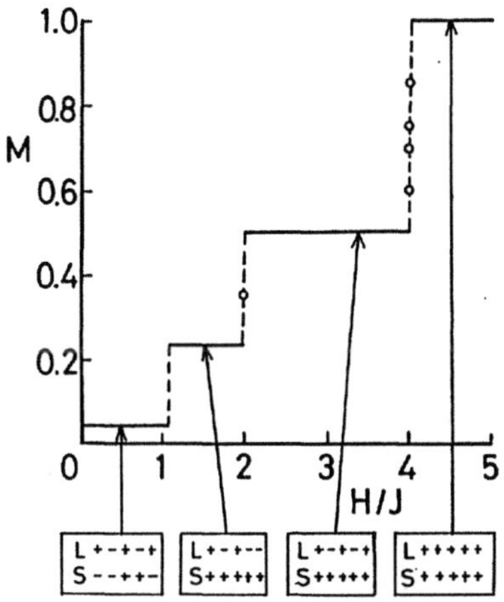

Fig. 6. Magnetization as a function of a uniform field for the antiferromagnetic Ising model on the dual 2D Penrose lattice. We only consider the states that all the spins on ten sublattices take the same orientation.

the magnetization as a function of an external magnetic field within the limits of this restricted states. Although we get a complicated step-wise structure, this is not the true ground state, of course. A hierarchical structure due to the self-similarity of quasicrystals may appear in the magnetization curve. The Monte Carlo study of this problem to get the phase diagram is interesting, and now in progress.

4. Summary

In summary, we have studied the critical phenomena of the Ising spins on the 3D icosahedral lattice by use of a Monte Carlo method. It has been shown that the critical exponents are universal among regular crystals and quasicrystals. We have also discussed the critical temperatures for the icoshedral quasilattice and its 'dual' lattice. The antiferromagnetic system on the dual Penrose lattice has been studied. We have found the growth of the sublattice magnetization at low temperatures. We may expect a similar behavior in the antiferromagnetic systems on the 'dual' icosahedral lattice.

References

[1] Y. Achiam, T. C. Lubensky and E. W. Marshall, Phys. Rev. **B33**, 6460 (1986).

[2] C. Godrèche, J.M. Luck and H. Orland, J. Stat. Phys. **45**, 777 (1986).
[3] T. C. Choy, Int. J. Mod. Phys. **B2**, 49 (1988).
[4] Y. Okabe and K. Niizeki, J. Phys. Soc. Jpn. **57** 16, 1536 (1988); J. Physique **49**, C8-1387 (1988).
[5] S. M. Bhattacharjee, J. S. Ho and J. A. Y. Johnson, J. Phys. **A20**, 4439 (1987).
[6] S. Miyajima, T. Odagaki and S. Miyashita, private communication.
[7] G. Amarendra, G. Ananthakrishna and G. Athithan, Europhys. Lett. **5**, 181 (1988).
[8] J. Oitmaa and M. J. Johnson, private communication.
[9] R. Abe and T. Dotera, J. Phys. Soc. Jpn. **58**, 3219 (1989).
[10] Y. Kinoshita and M. Suzuki, private communication.
[11] W. G. Wilson and C. A. Vause, Phys. Rev. **B39**, 689 (1989); **B39**, 4651 (1989).
[12] F. Yonezawa, S. Sakamoto and M. Hori, Phys. Rev. **B40**, 650 (1989).
[13] V. Elser and C. L. Henley, Phys. Rev. Lett. **55** 2883 (1985).
[14] G. Bhanot, D. Duke and R. Salvador, Phys. Rev. **B33**, 7841 (1986); J. Stat. Phys. **44**, 985 (1986).
[15] M. Kikuchi and Y. Okabe, Phys. Rev. **B35**, 5382 (1987); Prog. Theor. Phys. **78**, 540 (1987).
[16] M. N. Barber and W. Selke, J. Phys. **A15**, L617 (1982).
[17] I. Shozi, in *Phase Transitions and Critical Phenomena*, ed. C. Domb and M. S. Green (Academic Press, London) Vol. **1** (1972).

Index of Contributors

Springer Series in Solid-State Sciences

Editors: M. Cardona P. Fulde K. von Klitzing H.-J. Queisser

1 **Principles of Magnetic Resonance** 3rd Edition By C. P. Slichter
2 **Introduction to Solid-State Theory** 2nd Printing By O. Madelung
3 **Dynamical Scattering of X-Rays in Crystals** By Z. G. Pinsker
4 **Inelastic Electron Tunneling Spectroscopy** Editor: T. Wolfram
5 **Fundamentals of Crystal Growth I** Macroscopic Equilibrium and Transport Concepts. 2nd Printing By F. Rosenberger
6 **Magnetic Flux Structures in Superconductors** By R. P. Huebener
7 **Green's Functions in Quantum Physics** 2nd Edition By E. N. Economou
8 **Solitons and Condensed Matter Physics** 2nd Printing Editors: A. R. Bishop and T. Schneider
9 **Photoferroelectrics** By V. M. Fridkin
10 **Phonon Dispersion Relations in Insulators** By H. Bilz and W. Kress
11 **Electron Transport in Compound Semiconductors** By B. R. Nag
12 **The Physics of Elementary Excitations** By S. Nakajima, Y. Toyozawa, and R. Abe
13 **The Physics of Selenium and Tellurium** Editors: E. Gerlach and P. Grosse
14 **Magnetic Bubble Technology** 2nd Edition By A. H. Eschenfelder
15 **Modern Crystallography I** Symmetry of Crystals. Methods of Structural Crystallography By B. K. Vainshtein
16 **Organic Molecular Crystals** Their Electronic States. By E. A. Silinsh
17 **The Theory of Magnetism I** Statics and Dynamics. 2nd Printing By D. C. Mattis
18 **Relaxation of Elementary Excitations** Editors: R. Kubo and E. Hanamura
19 **Solitons.** Mathematical Methods for Physicists. 2nd Printing By G. Eilenberger
20 **Theory of Nonlinear Lattices** 2nd Edition By M. Toda
21 **Modern Crystallography II** Structure of Crystals By B. K. Vainshtein, V. M. Fridkin, and V. L. Indenbom
22 **Point Defects in Semiconductors I** Theoretical Aspects By M. Lannoo and J. Bourgoin
23 **Physics in One Dimension** Editors: J. Bernasconi, T. Schneider
24 **Physics in High Magnetic Fields** Editors: S. Chikazumi and N. Miura
25 **Fundamental Physics of Amorphous Semiconductors** Editor: F. Yonezawa
26 **Elastic Media with Microstructure I** One-Dimensional Models. By I. A. Kunin
27 **Superconductivity of Transition Metals** Their Alloys and Compounds By S. V. Vonsovsky, Yu. A. Izyumov, and E. Z. Kurmaev
28 **The Structure and Properties of Matter** Editor: T. Matsubara
29 **Electron Correlation and Magnetism in Narrow-Band Systems** Editor: T. Moriya
30 **Statistical Physics I** Equilibrium Statistical Mechanics By M. Toda, R. Kubo, N. Saito
31 **Statistical Physics II** Nonequilibrium Statistical Mechanics By R. Kubo, M. Toda, N. Hashitsume
32 **Quantum Theory of Magnetism** 2nd Edition By R. M. White
33 **Mixed Crystals** By A. I. Kitaigorodsky
34 **Phonons: Theory and Experiments I** Lattice Dynamics and Models of Interatomic Forces. By P. Brüesch
35 **Point Defects in Semiconductors II** Experimental Aspects By J. Bourgoin and M. Lannoo
36 **Modern Crystallography III** Crystal Growth 2nd Edition By A. A. Chernov
37 **Modern Crystallography IV** Physical Properties of Crystals Editor: L. A. Shuvalov
38 **Physics of Intercalation Compounds** Editors: L. Pietronero and E. Tosatti
39 **Anderson Localization** Editors: Y. Nagaoka and H. Fukuyama
40 **Semiconductor Physics** An Introduction 4th Edition By K. Seeger
41 **The LMTO Method** Muffin-Tin Orbitals and Electronic Structure By H. L. Skriver
42 **Crystal Optics with Spatial Dispersion, and Excitons** 2nd Edition By V. M. Agranovich and V. L. Ginzburg
43 **Structure Analysis of Point Defects in Solids by Multiple Magnetic Resonance Spectroscopy** By J.-M. Spaeth, J. R. Niklas, and R. H. Bartram
44 **Elastic Media with Microstructure II** Three-Dimensional Models By I. A. Kunin
45 **Electronic Properties of Doped Semiconductors** By B. I. Shklovskii and A. L. Efros
46 **Topological Disorder in Condensed Matter** Editors: F. Yonezawa and T. Ninomiya
47 **Statics and Dynamics of Nonlinear Systems** Editors: G. Benedek, H. Bilz, and R. Zeyher
48 **Magnetic Phase Transitions** Editors: M. Ausloos and R. J. Elliott
49 **Organic Molecular Aggregates** Electronic Excitation and Interaction Processes Editors: P. Reineker, H. Haken, and H. C. Wolf

Springer Series in Solid-State Sciences

Editors: M. Cardona P. Fulde K. von Klitzing H.-J. Queisser

50 **Multiple Diffraction of X-Rays in Crystals**
By Shih-Lin Chang

51 **Phonon Scattering in Condensed Matter**
Editors: W. Eisenmenger, K. Laßmann, and S. Döttinger

52 **Superconductivity in Magnetic and Exotic Materials** Editors: T. Matsubara and A. Kotani

53 **Two-Dimensional Systems, Heterostructures, and Superlattices**
Editors: G. Bauer, F. Kuchar, and H. Heinrich

54 **Magnetic Excitations and Fluctuations**
Editors: S. Lovesey, U. Balucani, F. Borsa, and V. Tognetti

55 **The Theory of Magnetism II** Thermodynamics and Statistical Mechanics By D. C. Mattis

56 **Spin Fluctuations in Itinerant Electron Magnetism** By T. Moriya

57 **Polycrystalline Semiconductors,**
Physical Properties and Applications
Editor: G. Harbeke

58 **The Recursion Method and Its Applications**
Editors: D. Pettifor and D. Weaire

59 **Dynamical Processes and Ordering on Solid Surfaces** Editors: A. Yoshimori and M. Tsukada

60 **Excitonic Processes in Solids**
By M. Ueta, H. Kanzaki, K. Kobayashi, Y. Toyozawa, and E. Hanamura

61 **Localization, Interaction, and Transport Phenomena** Editors: B. Kramer, G. Bergmann, and Y. Bruynseraede

62 **Theory of Heavy Fermions and Valence Fluctuations** Editors: T. Kasuya and T. Saso

63 **Electronic Properties of Polymers and Related Compounds**
Editors: H. Kuzmany, M. Mehring, and S. Roth

64 **Symmetries in Physics** Group Theory Applied to Physical Problems
By W. Ludwig and C. Falter

65 **Phonons: Theory and Experiments II**
Experiments and Interpretation of Experimental Results By P. Brüesch

66 **Phonons: Theory and Experiments III**
Phenomena Related to Phonons
By P. Brüesch

67 **Two-Dimensional Systems: Physics and New Devices**
Editors: G. Bauer, F. Kuchar, and H. Heinrich

68 **Phonon Scattering in Condensed Matter V**
Editors: A. C. Anderson and J. P. Wolfe

69 **Nonlinearity in Condensed Matter**
Editors: A. R. Bishop, D. K. Campbell, P. Kumar, and S. E. Trullinger

70 **From Hamiltonians to Phase Diagrams** The Electronic and Statistical-Mechanical Theory of sp-Bonded Metals and Alloys By J. Hafner

71 **High Magnetic Fields in Semiconductor Physics**
Editor: G. Landwehr

72 **One-Dimensional Conductors**
By S. Kagoshima, H. Nagasawa, and T. Sambongi

73 **Quantum Solid-State Physics**
Editors: S. V. Vonsovsky and M. I. Katsnelson

74 **Quantum Monte Carlo Methods** in Equilibrium and Nonequilibrium Systems Editor: M. Suzuki

75 **Electronic Structure and Optical Properties of Semiconductors** 2nd Edition
By M. L. Cohen and J. R. Chelikowsky

76 **Electronic Properties of Conjugated Polymers**
Editors: H. Kuzmany, M. Mehring, and S. Roth

77 **Fermi Surface Effects**
Editors: J. Kondo and A. Yoshimori

78 **Group Theory and Its Applications in Physics**
By T. Inui, Y. Tanabe, and Y. Onodera

79 **Elementary Excitations in Quantum Fluids**
Editors: K. Ohbayashi and M. Watabe

80 **Monte Carlo Simulation in Statistical Physics**
An Introduction
By K. Binder and D. W. Heermann

81 **Core-Level Spectroscopy in Condensed Systems**
Editors: J. Kanamori and A. Kotani

82 **Introduction to Photoemission Spectroscopy**
By S. Hüfner

83 **Physics and Technology of Submicron Structures**
Editors: H. Heinrich, G. Bauer, and F. Kuchar

84 **Beyond the Crystalline State** An Emerging Perspective By G. Venkataraman, D. Sahoo, and V. Balakrishnan

85 **The Fractional Quantum Hall Effect**
Properties of an Incompressible Quantum Fluid
By T. Chakraborty and P. Pietiläinen

86 **The Quantum Statistics of Dynamic Processes**
By E. Fick and G. Sauermann

87 **High Magnetic Fields in Semiconductor Physics II**
Transport and Optics Editor: G. Landwehr

88 **Organic Superconductors**
By T. Ishiguro and K. Yamaji

89 **Strong Correlation and Superconductivity**
Editors: H. Fukuyama, S. Maekawa, and A. P. Malozemoff

Springer Series in Solid-State Sciences

Batch number: 10620141

Printed by Printforce, the Netherlands